Ben Stacy Jerrik (Ed.)

Semiarundinaria

Ben Stacy Jerrik (Ed.)

Semiarundinaria

Bamboo, Bambusoideae

Part Press

Imprint

Permission is granted to copy, distribute and/or modify this document under the terms of the GNU Free Documentation License, Version 1.2 or any later version published by the Free Software Foundation; with no Invariant Sections, with the Front-Cover Texts, and with the Back- Cover Texts. A copy of the license is included in the section entitled "GNU Free Documentation License".

All parts of this book are extracted from Wikipedia, the free encyclopedia (www.wikipedia.org).

You can get detailed informations about the authors of this collection of articles at the end of this book. The editors (Ed.) of this book are no authors. They have not modified or extended the original texts.

Pictures published in this book can be under different licences than the GNU Free Documentation License. You can get detailed informations about the authors and licences of pictures at the end of this book.

The content of this book was generated collaboratively by volunteers. Please be advised that nothing found here has necessarily been reviewed by people with the expertise required to provide you with complete, accurate or reliable information. Some information in this book maybe misleading or wrong. The Publisher does not guarantee the validity of the information found here. If you need specific advice (f.e. in fields of medical, legal, financial, or risk management questions) please contact a professional who is licensed or knowledgeable in that area.

Any brand names and product names mentioned in this book are subject to trademark, brand or patent protection and are trademarks or registered trademarks of their respective holders. The use of brand names, product names, common names, trade names, product descriptions etc. even without a particular marking in this works is in no way to be construed to mean that such names may be regarded as unrestricted in respect of trademark and brand protection legislation and could thus be used by anyone.

Cover image: www.ingimage.com
Concerning the licence of the cover image please contact ingimage.

Publisher:
Part Press is a trademark of
International Book Market Service Ltd., 17 Rue Meldrum, Beau Bassin, 1713-01 Mauritius
Email: info@bookmarketservice.com
Website: www.bookmarketservice.com

Published in 2012

Printed in: U.S.A., U.K., Germany. This book was not produced in Mauritius.

ISBN: 978-613-8-75787-0

Contents

Semiarundinaria

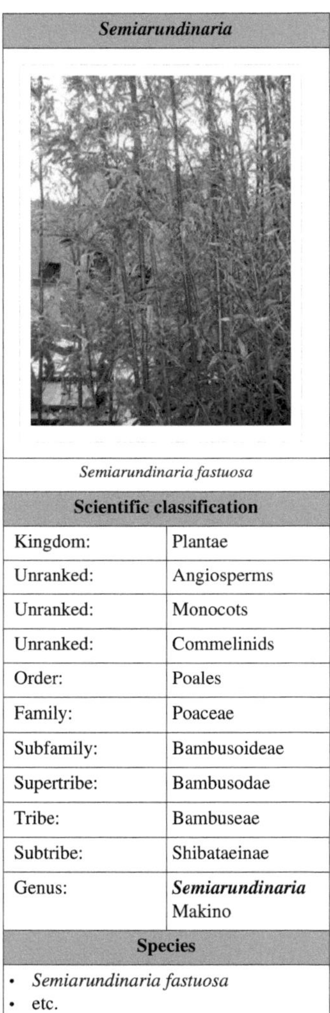

Semiarundinaria	
Semiarundinaria fastuosa	
Scientific classification	
Kingdom:	Plantae
Unranked:	Angiosperms
Unranked:	Monocots
Unranked:	Commelinids
Order:	Poales
Family:	Poaceae
Subfamily:	Bambusoideae
Supertribe:	Bambusodae
Tribe:	Bambuseae
Subtribe:	Shibataeinae
Genus:	***Semiarundinaria*** Makino
Species	

- *Semiarundinaria fastuosa*
- etc.

Semiarundinaria is a genus of tall or shrubby running bamboos. The species are found from temperate and subtropical regions of China and Japan.

See also

- *Brachystachyum*

Shibataeinae

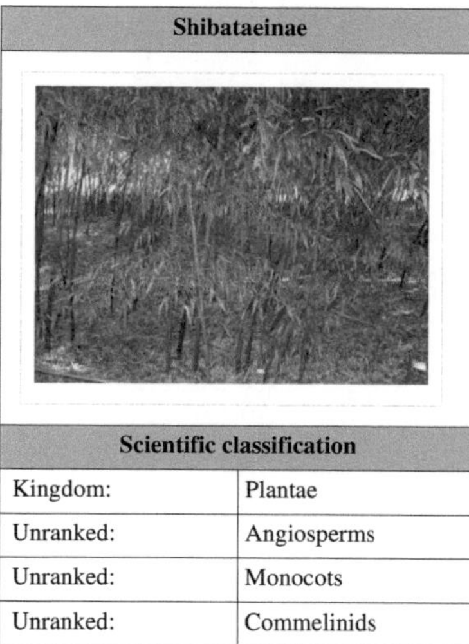

Shibataeinae	
Scientific classification	
Kingdom:	Plantae
Unranked:	Angiosperms
Unranked:	Monocots
Unranked:	Commelinids
Order:	Poales
Family:	Poaceae
Subfamily:	Bambusoideae
Supertribe:	Bambusodae
Tribe:	Bambuseae
Subtribe:	**Shibataeinae**
Genera	

- *Brachystachyum*
- *Chimonobambusa*
- *Indosasa*
- *Phyllostachys*
- *Qiongzhuea*
- *Semiarundinaria*
- *Shibataea*
- *Sinobambusa*
- *Temburongia* (incertae sedis)

Shibataeinae is a subtribe of bamboo (tribe Bambuseae of the family Poaceae). It comprises nine genera, though the position of *Temburongia* remains uncertain.

Bamboo

<table>
<tr><td colspan="2" align="center">Bamboo plant
Bambuseae tribe</td></tr>
<tr><td colspan="2" align="center"></td></tr>
<tr><td colspan="2" align="center">Bamboo forest in Kyoto, Japan</td></tr>
<tr><td colspan="2" align="center">Scientific classification</td></tr>
<tr><td>Kingdom:</td><td>Plantae</td></tr>
<tr><td>Unranked:</td><td>Angiosperms</td></tr>
<tr><td>Unranked:</td><td>Monocots</td></tr>
<tr><td>Unranked:</td><td>Commelinids</td></tr>
<tr><td>Order:</td><td>Poales</td></tr>
<tr><td>Family:</td><td>Poaceae</td></tr>
<tr><td>Subfamily:</td><td>Bambusoideae</td></tr>
<tr><td>Supertribe:</td><td>Bambusodae</td></tr>
<tr><td>Tribe:</td><td>Bambuseae
Kunth ex Dumort.</td></tr>
<tr><td colspan="2" align="center">Subtribes</td></tr>
<tr><td colspan="2">

- Arthrostylidiinae
- Arundinariinae
- Bambusinae
- Chusqueinae
- Guaduinae
- Melocanninae
- Nastinae
- Racemobambodinae
- Shibataeinae

See the full **Taxonomy of the Bambuseae**.
</td></tr>
<tr><td colspan="2" align="center">Diversity</td></tr>
<tr><td colspan="2" align="center">Around 92 genera and 5,000 species</td></tr>
</table>

Bamboo *listen(US)* is a group of perennial evergreens in the true grass family Poaceae, subfamily Bambusoideae, tribe **Bambuseae**. Giant bamboos are the largest members of the grass family. In bamboo, the internodal regions of the stem are hollow and the vascular bundles in the cross section are scattered throughout the stem instead of in a cylindrical arrangement. The dicotyledonous woody xylem is also absent. The absence of secondary growth wood causes the stems of monocots, even of palms and large bamboos, to be columnar rather than tapering.[1]

Bamboos are some of the fastest growing plants in the world,[2] as some species have been recorded as growing up to 100 cm (**unknown operator: u'strong'** in) within a 24 hour period due to a unique rhizome-dependent system. Bamboos are of notable economic and cultural significance in South Asia, South East Asia and East Asia, being used for building materials, as a food source, and as a versatile raw product.

Genus and geography

More than 70 genera are divided into about 1,450 species.[3] Bamboo species are found in diverse climates, from cold mountains to hot tropical regions. They occur across East Asia, from 50°N latitude in Sakhalin through to Northern Australia, and west to India and the Himalayas.[4] They also occur in sub-Saharan Africa, and in the Americas from the Mid-Atlantic United States[5] south to Argentina and Chile, reaching their southernmost point anywhere, at 47°S latitude. Continental Europe is not known to have any native species of bamboo.[6]

There have recently been some attempts to grow bamboo on a commercial basis in the Great Lakes region of eastern-central Africa, especially in Rwanda.[7] [8] Companies in the United States are growing, harvesting and distributing species such as Henon and Moso.[9]

Ecology

Growth

Bamboo is one of the fastest-growing plants on Earth with reported growth rates of 100 cm (**unknown operator: u'strong'** in) in 24 hours.[2] However, the growth rate is dependent on local soil and climatic conditions as well as species, and a more typical growth rate for many commonly cultivated bamboos in temperate climates is in the range of 3–10 cm (1-4 inches) per day during the growing period. Primarily growing in regions of warmer climates during the late Cretaceous period, vast fields existed in what is now Asia. Some of the largest timber bamboo can grow over 30 metres (**unknown operator: u'strong'** ft) tall, and be as large as 15–20 cm (6-8 inches) in diameter. However, the size range for mature bamboo is species dependent, with the smallest bamboos reaching only several inches high at maturity. A typical height range that would cover many of the common bamboos grown in the United States is 15–40 feet, depending on species.

Bamboo forest in Taiwan

Unlike trees, individual bamboo culms emerge from the ground at their full diameter and grow to their full height in a single growing season of 3–4 months. During these several months, each new shoot grows vertically into a culm with no branching out until the majority of the mature height is reached. Then the branches extend from the nodes and

Bamboo forest in Kwa-Zulu Natal

leafing out occurs. In the next year, the pulpy wall of each culm or stem slowly hardens. During the third year, the culm hardens further. The shoot is now considered a fully mature culm. Over the next 2–5 years (depending on species), fungus and mold begin to form on the outside of the culm, which eventually penetrate and overcome the culm. Around 5 − 8 years later (species and climate dependent), the fungal and mold growth cause the culm to collapse and decay. This brief life means culms are ready for harvest and suitable for use in construction within about 3 − 7 years. Individual bamboo culms do not get any taller or larger in diameter in subsequent years than they do in their first year, and they do not replace any growth that is lost from pruning or natural breakage. Bamboos have a wide range of hardiness depending on species and locale. Small or young specimens of an individual species will produce small culms initially. As the clump and its rhizome system matures, taller and larger culms will be produced each year until the plant approaches its particular species limits of height and diameter.

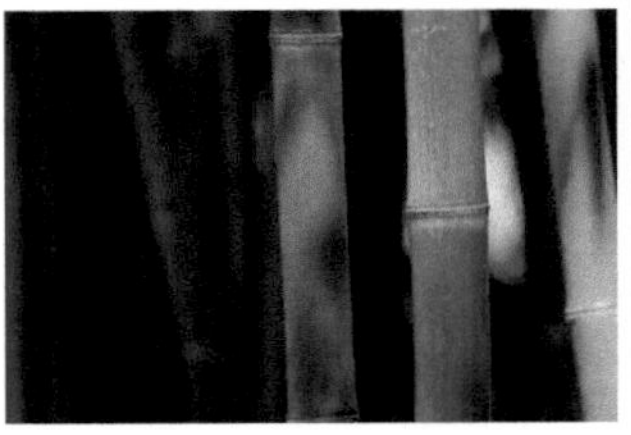
Close up of bamboo stalk

Many tropical bamboo species will die at or near freezing temperatures, while some of the hardier or so-called temperate bamboos can survive temperatures as low as −29 °C (−**unknown operator: u'strong'** °F). Some of the hardiest bamboo species can be grown in places as cold as USDA Plant Hardiness Zones 5-6, although they typically will defoliate and may even lose all above-ground growth; yet the rhizomes will survive and send up shoots again the next spring. In milder climates, such as USDA Zone 8 and above, some hardy bamboo may remain fully leafed out year around.

Mass flowering

Most bamboo species flower infrequently. In fact, many bamboos only flower at intervals as long as 65 or 120 years. These taxa exhibit mass flowering (or gregarious flowering), with all plants in a particular species flowering worldwide over a several year period. The longest mass flowering interval known is 130 years, and is found for all the species *Phyllostachys bambusoides* (Sieb. & Zucc.). In this species, all plants of the same stock flower at the same time, regardless of differences in geographic locations or climatic conditions, and then the bamboo dies. The lack of environmental impact on the time of flowering indicates the presence of some sort of "alarm clock" in each cell of the plant which signals the diversion of all energy to flower production and the cessation of vegetative growth.[10] This mechanism, as well as the evolutionary cause behind it, is still largely a mystery.

Flowering bamboo

One theory to explain the evolution of this semelparous mass flowering is the predator satiation hypothesis. This theory argues that by fruiting at the same time, a population increases the survival rate of their seeds by flooding the area with fruit so that even if predators eat their fill, there will still be seeds left over. By having a flowering cycle longer than the lifespan of the rodent predators, bamboos can regulate animal populations by causing starvation during the period between flowering events. Thus, according to this hypothesis, the death of the adult clone is due to resource exhaustion, as it would be more effective for parent plants to devote all resources to creating a large seed crop than to hold back energy for their own regeneration.[11]

A second theory, the fire cycle hypothesis, argues that periodic flowering followed by death of the adult plants has evolved as a mechanism to create disturbance in the habitat, thus providing the seedlings with a gap in which to grow. This hypothesis argues that the dead culms create a large fuel load, and also a large target for lightning strikes, increasing the likelihood of wildfire.[12] Because bamboos can be aggressive as early successional plants, the seedlings would be able to outstrip other plants and take over the space left by their parents.

However, both have been disputed for different reasons. The predator satiation theory does not explain why the flowering cycle is 10 times longer than the lifespan of the local rodents, something not predicted by the theory. The bamboo fire cycle theory is considered by a few scientists to be unreasonable; they argue[13] that fires only result from humans and there is no natural fire in India. This notion is considered wrong based on distribution of lightning strike data during the dry season throughout India. However, another argument against this theory is the lack of precedent for any living organism to harness something as unpredictable as lightning strikes to increase its chance of survival as part of natural evolutionary progress.[14]

The mass fruiting also has direct economic and ecological consequences, however. The huge increase in available fruit in the forests often causes a boom in rodent populations, leading to increases in disease and famine in nearby human populations. For example, there are devastating consequences when the *Melocanna bambusoides* population flowers and fruits once every 30–35 years [15] around the Bay of Bengal. The death of the bamboo plants following their fruiting means the local people lose their building material, and the large increase in bamboo fruit leads to a rapid increase in rodent populations. As the number of rodents increase, they consume all available food, including grain fields and stored food, sometimes leading to famine. These rats can also carry dangerous diseases such as typhus, typhoid, and bubonic plague, which can reach epidemic proportions as the rodents increase in number.[10] [11] The relationship between rat populations and bamboo flowering was examined in a 2009 Nova documentary Rat Attack.

In any case, flowering produce masses of seeds, typically suspended from the ends of the branches. These seeds will give rise to a new generation of plants that may be identical in appearance to those that preceded the flowering, or they may also produce new cultivars with different characteristics, such as the presence or absence of striping or other changes in coloration of the culms.

Bamboo in animal diets

Soft bamboo shoots, stems, and leaves are the major food source of the giant panda of China, the red panda of Nepal and the bamboo lemurs of Madagascar. Rats will eat the fruits as described above. Mountain gorillas of Africa also feed on bamboo, and have been documented consuming bamboo sap which was fermented and alcoholic;[16] chimps and elephants of the region also eat the stalks.

Bamboo is the main food of the giant panda; it makes up 99% of the panda's diet.

Cultivation

Commercial timber

Timber is harvested from cultivated and wild stands and some of the larger bamboos, particularly species in the genus *Phyllostachys*, are known as "timber bamboos".

Harvesting

Bamboo used for construction purposes must be harvested when the culms reach their greatest strength and when sugar levels in the sap are at their lowest, as high sugar content increases the ease and rate of pest infestation.

Harvesting of bamboo is typically undertaken according to the following cycles:

1) Life cycle of the culm: As each individual culm goes through a 5–7 year life cycle, culms are ideally allowed to reach this level of maturity prior to full capacity harvesting. The clearing out or thinning of culms, particularly older decaying culms, helps to ensure adequate light and resources for new growth. Well-maintained clumps may have a productivity three to four times that of an unharvested wild clump.

Bamboo foliage with yellow stems
(probably *Phyllostachys aurea*)

2) Life cycle of the culm: Consistent with the life cycle described above, bamboo is harvested from two to three years through to five to seven years, depending on the species.

3) Annual cycle: As all growth of new bamboo occurs during the wet season, disturbing the clump during this phase will potentially damage the upcoming crop. Also during this high rain fall period, sap levels are at their highest, and then diminish towards the dry season. Picking immediately prior to the wet/growth season may also damage new shoots. Hence, harvesting is best at the end of the dry season, a few months prior to the start of the wet.

Bamboo foliage with black stems
(probably *Phyllostachys nigra*)

4) Daily cycle: During the height of the day, photosynthesis is at its peak, producing the highest levels of sugar in sap, making this the least ideal time of day to harvest. Many traditional practitioners believe the best time to harvest is at dawn or dusk on a waning moon. This practice makes sense in terms of both moon cycles, visibility and daily cycles.

Leaching

Leaching is the removal of sap after harvest. In many areas of the world, the sap levels in harvested bamboo are reduced either through leaching or postharvest photosynthesis. Examples of this practice include:

1. Cut bamboo is raised clear of the ground and leant against the rest of the clump for one to two weeks until leaves turn yellow to allow full consumption of sugars by the plant.
2. A similar method is undertaken, but with the base of the culm standing in fresh water, either in a large drum or stream to leach out sap.
3. Cut culms are immersed in a running stream and weighted down for three to four weeks.
4. Water is pumped through the freshly cut culms, forcing out the sap (this method is often used in conjunction with the injection of some form of treatment).

In the process of water leaching, the bamboo is dried slowly and evenly in the shade to avoid cracking in the outer skin of the bamboo, thereby reducing opportunities for pest infestation.

Durability of bamboo in construction is directly related to how well it is handled from the moment of planting through harvesting, transportation, storage, design, construction and maintenance. Bamboo harvested at the correct time of year and then exposed to ground contact or rain, will break down just as quickly as incorrectly harvested material.

Ornamental bamboos

There are two general patterns for the growth of bamboo: "clumping" (sympodial) and "running" (monopodial). Clumping bamboo species tend to spread slowly, as the growth pattern of the rhizomes is to simply expand the root mass gradually, similar to ornamental grasses. "Running" bamboos, on the other hand, need to be taken care of in cultivation because of their potential for aggressive behavior. They spread mainly through their roots and/or rhizomes, which can spread widely underground and send up new culms to break through the surface. Running bamboo species are highly variable in their tendency to spread; this is related to both the species and the soil and climate conditions. Some can send out runners of several meters a year, while others can stay in the same general area for long periods. If neglected, over time they can cause problems by moving into adjacent areas.

Bamboos seldom and unpredictably flower, and the frequency of flowering varies greatly from species to species. Once flowering takes place, a plant will decline and often die entirely. Although there are always a few species of bamboo in flower at any given time, collectors desiring to grow specific bamboo typically obtain their plants as divisions of already-growing plants, rather than waiting for seeds to be produced.

Regular maintenance will indicate major growth directions and locations. Once the rhizomes are cut, they are typically removed; however, rhizomes take a number of months to mature and an immature, severed rhizome will usually cease growing if left in-ground. If any bamboo shoots come up outside of the bamboo area afterwards, their presence indicates the precise location of the missed rhizome. The fibrous roots that radiate from the rhizomes do not produce more bamboo if they stay in the ground.

Bamboo growth can also be controlled by surrounding the plant or grove with a physical barrier. Typically, concrete and specially-rolled HDPE plastic are the materials used to create the barrier, which is placed in a 60–90 cm (**unknown operator: u'strong'unknown operator: u'strong'unknown operator: u'strong' unknown operator: u'strong'**) deep ditch around the planting, and angled out at the top to direct the rhizomes to the surface. (This is only possible if the barrier is installed in a straight line.) If the containment area is small, this method can be detrimental to ornamental bamboo as the bamboo within can become rootbound and start to display the signs of any unhealthy containerized plant. In addition, rhizomes can escape over the top, or beneath the barrier if it is not deep enough. Strong rhizomes and tools can penetrate plastic barriers, so care must be taken. In small areas, regular maintenance may be the best method for controlling the running bamboos. Barriers and edging are unnecessary for clump-forming bamboos, although clump-forming bamboos may eventually need to have portions removed if they become too large.

The ornamental plant sold in containers and marketed as "lucky bamboo" is actually an entirely unrelated plant, *Dracaena sanderiana*. It is a resilient member of the lily family that grows in the dark, tropical rainforests of Southeast Asia and Africa. Lucky bamboo has long been associated with the Eastern practice of *feng shui*. On a similar note, Japanese knotweed is also sometimes mistaken for a bamboo, but it grows wild and is considered an invasive species.

Uses

Culinary

The shoots (new culms that come out of the ground) of bamboo are edible. They are used in numerous Asian dishes and broths, and are available in supermarkets in various sliced forms, in both fresh and canned versions. The shoots of the giant bamboo (*Cathariostachys madagascariensis*) contain cyanide. Despite this, the golden bamboo lemur ingests many times the quantity of toxin that would kill a human.

Edible bamboo shoots in a Japanese market

The bamboo shoot in its fermented state forms an important ingredient in cuisines across the Himalayas. In Assam, India, for example, it is called *khorisa*. In Nepal, a delicacy popular across ethnic boundaries consists of bamboo shoots fermented with turmeric and oil, and cooked with potatoes into a dish that usually accompanies rice (*alu tama* in Nepali).

In Indonesia, they are sliced thin and then boiled with *santan* (thick coconut milk) and spices to make a dish called *gulai rebung*. Other recipes using bamboo shoots are *sayur lodeh* (mixed vegetables in coconut milk) and *lun pia* (sometimes written *lumpia*: fried wrapped bamboo shoots with vegetables). The shoots of some species contain toxins that need to be leached or boiled out before they can be eaten safely.

Khao lam (Thai: ข้าวหลาม) is glutinous rice with sugar and coconut cream cooked in specially-prepared bamboo sections of different diameters and lengths

Pickled bamboo, used as a condiment, may also be made from the pith of the young shoots.

The sap of young stalks tapped during the rainy season may be fermented to make *ulanzi* (a sweet wine) or simply made into a soft drink. Bamboo leaves are also used as wrappers for steamed dumplings which usually contains glutinous rice and other ingredients.

Pickled bamboo shoots (Nepali: tama) are cooked with black eyed beans as a delicacy food in Nepal. Many Nepalese restaurant around the world serve this dish as *aloo bodi tama*. Fresh bamboo shoots are sliced and pickled with mustard seeds and turmeric and kept in glass jar in sun for the best taste. It is used alongside many dried beans in cooking during winter months. Baby shoots (Nepali: tusa) of a very different variety of bamboo (Nepali: Nigalo) native to Nepal is cooked as a curry in Hilly regions.

In Sambalpur, India, the tender shoots are grated into juliennes and fermented to prepare *kardi*. The name is derived from the Sanskrit word for bamboo shoot, *karira*. This fermented bamboo shoot is used in various culinary preparations, notably *amil*, a sour vegetable soup. It is also made into pancakes using rice flour as a binding agent. The shoots that have turned a little fibrous are fermented, dried, and ground to sand-sized particles to prepare a garnish known as *hendua*. It is also cooked with tender pumpkin leaves to make sag green leaves.

The empty hollow in the stalks of larger bamboo is often used to cook food in many Asian cultures. Soups are boiled and rice is cooked in the hollows of fresh stalks of bamboo directly over a flame. Similarly, steamed tea is sometimes rammed into bamboo hollows to produce compressed forms of Pu-erh tea. Cooking food in bamboo is said to give the food a subtle but distinctive taste.

In addition, bamboo is frequently used for cooking utensils within many cultures, and is used in the manufacture of chopsticks. In modern times, some see bamboo tools as an eco-friendly alternative to other manufactured utensils.

Medicine

Bamboo is used in Chinese medicine for treating infections and healing.

It is a low-calorie source of potassium. It is known for its sweet taste and as a good source of nutrients and protein.

In Ayurveda, the Indian system of traditional medicine, the silicious concretion found in the culms of the bamboo stem is called *banslochan*. It is known as *tabashir* or *tawashir* in *Unani-Tibb* the Indo-Persian system of medicine. In English, it is called "bamboo manna". This concretion is said to be a tonic for the respiratory diseases. It was earlier obtained from *Melocanna bambusoides* and is very hard to get. In most Indian literature, *Bambusa arundinacea* is described as the source of bamboo manna.[17]

Construction

In its natural form, bamboo as a construction material is traditionally associated with the cultures of South Asia, East Asia and the South Pacific, to some extent in Central and South America and by extension in the aesthetic of Tiki culture. In China and India, bamboo was used to hold up simple suspension bridges, either by making cables of split bamboo or twisting whole culms of sufficiently pliable bamboo together. One such bridge in the area of Qian-Xian is referenced in writings dating back 960 A.D. and may have stood since as far back as the 3rd century B.C., due largely to continuous maintenance.[18]

House made entirely of bamboo

Bamboo has also long been used as scaffolding; the practice has been banned in China for buildings over six storeys but is still in continuous use for skyscrapers in Hong Kong.[19] In the Philippines, the nipa hut is a fairly typical example of the most basic sort of housing where bamboo is used; the walls are split and woven bamboo, and bamboo slats and poles may be used as its support. In Japanese architecture, bamboo is used primarily as a supplemental and/or decorative element in buildings such as fencing, fountains, grates and gutters, largely due to the ready abundance of quality timber.[20]

Bamboo scaffolding can reach great heights.

Various structural shapes may be made by training the bamboo to assume them as it grows. Squared sections of bamboo are created by compressing the growing stalk within a square form.[21] Arches may similarly be created by forcing the bamboo's growth with the desired form, and costs much less than it would to assume the same shape in regular wood timber. More traditional forming methods, such as the application of heat and pressure, may also be used to curve or flatten the cut stalks.[22]

Bamboo can be cut and laminated into sheets and planks. This process involves cutting stalks into thin strips, planing them flat, boiling and drying the strips, which are then glued, pressed and finished.[23] Generally long used in China and Japan, entrepreneurs started developing and selling laminated bamboo flooring in the West during the mid 1990s;[23] products made from bamboo laminate, including flooring, cabinetry, furniture and even decorations, are currently surging in popularity, transitioning from the boutique market to mainstream providers such as Home Depot. The bamboo goods industry (which also includes small goods, fabric, etc.) is expected to be worth $25 billion by the year 2012.[24] The quality of bamboo laminate varies between manufacturers and the maturity of the plant from which it was harvested (six years being considered the optimum); the sturdiest products fulfill their claims of being up to three times harder than oak hardwood, but others may be softer than standard hardwood.[23]

Bamboo intended for use in construction should be treated to resist insects and rot. The most common solution for this purpose is a mixture of borax and boric acid.[25] Another process involves boiling cut bamboo to remove the starches that attract insects.[23]

Bamboo has been used as reinforcement for concrete in those areas where it is plentiful, though dispute exists over its effectiveness in the various studies done on the subject. Bamboo does have the necessary strength to fulfill this function, but untreated bamboo will swell from the absorption of water from the concrete, causing it to crack. Several procedures must be followed to overcome this shortcoming.[26]

Several institutes, businesses, and universities are working on the bamboo as an ecological construction material. In the United States and France, it is possible to get houses made entirely of

Bamboo pavilion in the Shenzhen Biennale

bamboo, which are earthquake and cyclone-resistant and internationally certified. In Bali, Indonesia, an international primary school, named the Green School [27], is constructed entirely of bamboo, due to its beauty, and advantages as a sustainable resource. There are three ISO standards for bamboo as a construction material.

In parts of India, bamboo is used for drying clothes indoors, both as the rod high up near the ceiling to hang clothes on, as well as the stick that is wielded with acquired expert skill to hoist, spread, and to take down the clothes when dry. It is also commonly used to make ladders, which apart from their normal function, are also used for carrying bodies in funerals. In Maharashtra, the bamboo groves and forests are called VeLuvana, the name *VeLu* for bamboo is most likely from Sanskrit, while *Vana* means forest.

Furthermore, bamboo is also used to create flagpoles for saffron-coloured, Hindu religious flags, which can be seen fluttering across India, especially Bihar and Uttar Pradesh, as well as in Guyana and Suriname.

Bamboo is used for the structural members of the India pavilion at Expo 2010 in Shanghai. The pavilion is the world's largest bamboo dome, about 34 m in diameter, with bamboo beams/members overlaid with a ferro-cement slab, water proofing, copper plate, solar PV panels, a small windmill and live plants. A total of 30 km of bamboo was used. The dome is supported on 18-m-long steel piles and a series of steel ring beams. The bamboo was treated with borax and boric acid as a fire retardant and insecticide and bent in the required shape. The bamboo sections are joined with reinforcement bars and concrete mortar to achieve necessary lengths.[28]

Textiles

Because the fibers of bamboo are very short (less than 3mm), they are impossible to transform into yarn in a natural process.[29] The usual process by which textiles labeled as being made of bamboo are produced uses only the rayon, that is being made out of the fibers with heavy employment of chemicals. To accomplish this, the fibers are broken down with chemicals and extruded through mechanical spinnerets; the chemicals include lye, carbon disulfide and strong acids.[23] Retailers have sold both end products as "bamboo fabric" to cash in on bamboo's current ecofriendly cachet; however, the Canadian Competition Bureau[30] and the US Federal Trade Commission,[31] as of mid-2009, are cracking down on the practice of labeling bamboo rayon as natural bamboo fabric. Under the guidelines of both agencies, these products must be labeled as rayon with the optional qualifier "from bamboo".[31]

Paper

Bamboo fiber has been used to make paper in China since early times. A high quality hand-made paper is still produced in small quantities. Coarse bamboo paper is still used to make spirit money in many Chinese communities.[32]

Bamboo pulps are mainly produced in China, Myanmar, Thailand and India and are used in printing and writing papers.[33] The most common bamboo species used for paper are *Dendrocalamus asper* and *Bamboo bluemanea*. It is also possible to make dissolving pulp from bamboo. The average fiber length is similar to hardwoods, but the properties of bamboo pulp are closer to softwoods pulps due to it having a very broad fiber length distribution.[33] With the help of molecular tools, it is now possible to distinguish the superior fiber-yielding species/varieties even at juvenile stages of their growth which can help in unadulterated merchandise production.[34]

Musical instruments

Bamboo's natural hollow form makes it an obvious choice for many instruments, particularly wind and percussion. There are numerous types of bamboo flute made all over the world, such as the *dizi, xiao, shakuhachi, palendag* and *jinghu*. In India, it is a very popular and highly respected musical instrument, available even to the poorest and the choice of many highly venerated maestros of classical music. It is known and revered above all as the divine flute forever associated with Lord Krishna, who is always portrayed holding a *bansuri* in sculptures and paintings. Four of the instruments used in Polynesia for traditional hula are made of bamboo: nose flute, rattle, stamping pipes and the jaw harp. Bamboo may be used in the construction of the Australian didgeridoo instead of the more traditional eucalyptus wood. In Indonesia and the Philippines, bamboo has been used for making various kinds of musical instruments, including the *kolintang, angklung* and *bumbong*. Traditional Philippine *banda kawayan* (bamboo bands) use a variety of bamboo musical instruments, including the marimba, *angklung*, panpipes and *bumbong*, as well as bamboo versions of western instruments, such as clarinets, saxophones, and tubas.[35] The Las Piñas Bamboo Organ in the Philippines has pipes made of bamboo culms. The modern amplified string instrument, the Chapman stick, is also constructed using bamboo. The *khene* (also spelled *khaen, kaen* and *khen*; Lao: , Thai: แคน) is a mouth organ of Lao origin whose pipes, which are usually made of bamboo, are connected with a small, hollowed-out hardwood reservoir into which air is blown, creating a sound similar to that of the violin. In the Indian Ocean island of Madagascar, the *valiha*, a long tube zither made of a single bamboo stalk, is considered the national instrument. Bamboo has also recently been used for the manufacture of Ukuleles. Bamboo Ukuleles are constructed of solid cross laminated bamboo strips not plywood. The bamboo solid wood strips are similar to bamboo manufactured flooring. In addition to their strength, bamboo ukuleles have excellent sound & rival ukuleles made out of more traditional woods like Mahogany and KOA. Bamboo makes an excellent choice for an eco-friendly cost conscious ukulele aficionados.

Other Uses

Bamboo has a long history of use in Asian furniture. Chinese bamboo furniture is a distinct style based on millennia-long tradition.

Several manufacturers offer bamboo bicycles and longboards.[36]

Due to its flexibility bamboo is also used to make fishing rods. The split cane rod is especially prized for fly fishing. Bamboo has been traditionally used in Malaysia as a firecracker called a meriam buluh or bamboo cannon. Four-foot long sections of bamboo are cut, and a mixture of water and calcium carbide are introduced. The resulting acetylene gas is ignited with a stick producing a loud bang. Bamboo can be used in water desalination. A bamboo filter is used to remove the salt from seawater.[37]

In Asian culture

Bamboo's long life makes it a Chinese symbol of longevity, while in India it is a symbol of friendship. The rarity of its blossoming has led to the flowers' being regarded as a sign of impending famine. This may be due to rats feeding upon the profusion of flowers, then multiplying and destroying a large part of the local food supply. The most recent flowering began in May 2006 (see Mautam). Bamboo is said to bloom in this manner only about every 50 years (see 28–60 year examples in FAO: 'gregarious' species table [38]).

Bamboo, by Xu Wei, Ming Dynasty.

In Chinese culture, the bamboo, plum blossom, orchid, and chrysanthemum (often known as *méi lán zhú jú*) are collectively referred to as the Four Gentlemen. These four plants also represent the four seasons and, in Confucian ideology, four aspects of the *junzi* ("prince" or "noble one"). The pine (*sōng*), the bamboo (*zhú*), and the plum blossom (*méi*) are also admired for their perseverance under harsh conditions, and are together known as the "Three Friends of Winter" (*suìhán sānyǒu*) in Chinese culture. The "Three Friends of Winter" is traditionally used as a system of ranking in Japan, for example in sushi sets or accommodations at a traditional *ryokan*. Pine (*matsu*) is of the first rank, bamboo (*také*) is of second rank, and plum (*ume*) is of the third.

Bamboo, noble and useful

Bamboo, one of the "four gentlemen" (bamboo, orchid, plum blossom and chrysanthemum), plays such an important role in traditional Chinese culture that it is even regarded as a behaviour model of the gentleman. As bamboo has some features like upright, tenacity and hollow heart, people endow bamboo with integrity, elegance and plainness though it is not physically strong. Ancient Chinese poets wrote countless poems to praise bamboo, but actually they were truly talking about people like bamboo and express their understanding of what is a real gentleman should be like. According to Laws,an ancient poet Bai, Juyi (772-846) thought that to be a gentleman, a man doesn't need to be physically strong, but he must be mentally strong. He must be upright, perseverant. Just as a bamboo is hollow-hearted, he should open his heart to accept anything of benefit and never have arrogance or prejudice. Bamboo is not only a symbol of gentleman, but also an important role in Buddhism. In the first century, Buddhism was introduced into China. As cannons of Buddhism don't allow its believers to do anything cruel to animals, meat, egg and fish were not allowed in the diet. However, people need something nutritional to live, thus, the tender bamboo shoot (it is called "*sǔn*" in Chinese) became a good choice. The bamboo shoot is nutritional and eating it does not violate the cannon. With thousands of years' development, how to eat bamboo shoot has become a part of cuisine system, especially for monks. A Buddhist monk named Zan Ning, wrote a manual of the bamboo shoot called "*Sǔn Pǔ*". He offered descriptions and recipes for many kinds of bamboo shoots.[39] Bamboo shoot has always been a traditional dish on Chinese's dinner table, especially in southern China. In ancient time, as long as people have money to buy a big house with yard, they will always plant bamboos in their garden. Bamboo is a necessary element of Chinese culture, or even in the whole Asian civilization. People plant bamboos, eat bamboo shoots, paint bamboos, write poems for bamboos, and speak highly of gentlemen who are like bamboos. Bamboo, is not only a plant, but also a part of people's life.

In Japan, a bamboo forest sometimes surrounds a Shinto shrine as part of a sacred barrier against evil. Many Buddhist temples also have bamboo groves.

In northern Indian state of Assam, the fermented bamboo paste known as *khorisa* is known locally as a folk remedy for the treatment of impotence, infertility, and menstrual pains.

Bamboo plays an important part of the culture of Vietnam. Bamboo symbolizes the spirit of Vovinam (a Vietnamese martial arts): *cường nhu phối triển* (coordination between hard and soft (martial arts)). Bamboo also symbolizes the Vietnamese hometown and Vietnamese soul: the gentlemanlike, straightforwardness, hard working, optimism, unity and adaptability. A Vietnamese proverb says: "When the bamboo is old, the bamboo sprouts appear", the meaning being Vietnam will never be annihilated; if the previous generation dies, the children take their place. Therefore, the Vietnam nation and Vietnamese value will be maintained and developed eternally. Traditional Vietnamese villages are surrounded by thick bamboo hedges (*lũy tre*).

The Song Dynasty (960-1279 AD) Chinese scientist and polymath Shen Kuo (1031–1095) used the evidence of underground petrified bamboo found in the dry northern climate of Yan'an, Shanbei region, Shaanxi province to support his geological theory of gradual climate change.[40] [41]

A cylindrical bamboo brush holder or holder of poems on scrolls, created by Zhang Xihuang in the 17th century, late Ming or early Qing Dynasty - in the calligraphy of Zhang's style, the poem *Returning to My Farm in the Field* by the fourth century poet Tao Yuanming is incised on the holder.

Myths and legends

Several Asian cultures, including that of the Andaman Islands, believe humanity emerged from a bamboo stem. In the Philippine creation myth, legend tells that the first man and the first woman each emerged from split bamboo stems on an island created after the battle of the elemental forces (Sky and Ocean). In Malaysian legends a similar story includes a man who dreams of a beautiful woman while sleeping under a bamboo plant; he wakes up and breaks the bamboo stem, discovering the woman inside. The Japanese folktale "Tale of the Bamboo Cutter"

Bamboo-style barred window in Lin An Tai Historical House, Taipei

(*Taketori Monogatari*) tells of a princess from the Moon emerging from a shining bamboo section. Hawaiian bamboo ('ohe) is a *kinolau* or body form of the Polynesian creator god Kāne.

Bamboo cane is also the weapon of Vietnamese legendary hero Saint Giong - who had grown up immediately and magically since the age of three because of his national liberating wish against Ân invaders.

An ancient Vietnamese legend (The Hundred-knot Bamboo Tree) tells of a poor, young farmer who fell in love with his landlord's beautiful daughter. The farmer asked the landlord for his daughter's hand in marriage, but the proud landlord would not allow her to be bound in marriage to a poor farmer. The landlord decided to foil the marriage with an impossible deal; the farmer must bring him a "bamboo tree of one-hundred nodes". But Buddha (*Bụt*) appeared to the farmer and told him that such a tree could be made from one-hundred nodes from several different trees. *Bụt* gave to him four magic words to attach the many nodes of bamboo: *Khắc nhập, khắc xuất*, which means "joined together immediately, fell apart immediately". The triumphant farmer returned to the landlord and demanded his daughter. Curious to see such a long bamboo, the landlord was magically joined to the bamboo when he touched it as the young farmer said the first two magic words. The story ends with the happy marriage of the farmer and the landlord's daughter after the landlord agreed to the marriage and asked to be separated from the bamboo.

In a Chinese legend, the Emperor Yao gave two of his daughters as a test for his potential to rule to the future Emperor Shun. Shun passed the test of being able to run his household with the two emperor's daughters as wives, and thus Yao made Shun his successors, bypassing his unworthy son. Later, Shun drowned in the Xiang River. The tears his two bereaved wives let fall upon the bamboos growing there explains the origin of spotted bamboo. The two women later became goddesses.

As a writing surface

Bamboo was in widespread use in early China as a medium for written documents. The earliest surviving examples of such documents, written in ink on string-bound bundles of bamboo strips (or "slips"), date from the fifth century BC during the Warring States period. However, references in earlier texts surviving on other media make it clear that some precursor of these Warring States period bamboo slips was in use as early as the late Shang period (from about 1250 BC).

Bamboo or wooden strips were the standard writing material during the Han dynasty, and excavated examples have been found in abundance.[42] Subsequently, paper began to displace bamboo and wooden strips from mainstream uses, and by the fourth century AD, bamboo slips had been largely abandoned as a medium for writing in China. Several paper industries are surviving on bamboo forests. Ballarpur (Chandrapur, Maharstra) paper mills use bamboo for paper production.

As a weapon

Bamboo is used in several East Asian and South Asian martial arts.

- In the ancient Tamil martial art of Silambam, fighters would hit each other rapidly with bamboo sticks.
- In the Japanese martial art Kendo, bamboo is used to make the Shinai sword.
- A bamboo stick can be made into a simple spear by sharpening one of the ends
- Archery longbow and recurve bow limbs are commonly crafted with flat ground bamboo, and make superior weapons for bowhunting and target archery.
- History's first gunpowder-based weapons, such as the Fire lance, were made of bamboo.

Other cultures

The ethnic group known as the Bozo of West Africa, take their name from the Bambara phrase *bo-so*, which means "bamboo house".

Bamboo is the national plant of St. Lucia.

See also

- Bamboo Farm and Coastal Gardens
- *Menstruocalamus*
- Plant textiles
- International Network for Bamboo and Rattan
- Big Bambú
- Bamboo Processing Machine
- Xiang River goddesses

References

[1] Botany; Wilson,C.L. and Loomis,W.E. Third edition. Holt, Rinehart and Winston
[2] Farrelly, David (1984). *The Book of Bamboo*. Sierra Club Books. ISBN 0-87156-825-X.
[3] Gratani, Loretta; Maria Fiore Crescente, Laura Varone, Giuseppe Fabrini, and Eleonora Digiulio (2008). "Growth pattern and photosynthetic activity of different bamboo species growing in the Botanical Garden of Rome". *Flora* **203**: 77–84..
[4] Bystriakova, N.; N. Bystriakova, V. Kapos, I. Lysenko and C.M.A. Stapleton (September 2003). "Distribution and conservation status of forest bamboo biodiversity in the Asia-Pacific Region" (http://www.springerlink.com/content/gu726j88x87k4508/). *Biodiversity and Conservation* **12** (9): 1833–1841. doi:10.1023/A:1024139813651. . Retrieved 2009-08-12.
[5] "Arundinaria gigantea (Walt.) Muhl. giant cane" (http://plants.usda.gov/java/profile?symbol=ARGI). *PLANTS Database*. USDA. .
[6] editor-in-chief, Anthony Huxley, editor, Mark Griffiths, managing editor, Margot Levy. (1992). Huxley, A.. ed. *New RHS Dictionary of Gardening*. Macmillan New RHS Dictionary of Gardening. ISBN 0-333-47494-5.

[7] "BAMBOO FARMING: AN OPPORTUNITY TO TRANSFORM LIVELIHOODS," [[The New Times (Rwanda)|The Sunday Times (http://www.newtimes.co.rw/index.php?issue=14283&article=7040)], June 6, 2010]

[8] "Cash in on Bamboo farming, Bazivamo urges farmers," Stevenson Mugisha, The New Times, June 7, 2010 (http://www.newtimes.co.rw/index.php?issue=14282&article=29953)

[9] McDill, Stephen. "MS Business Journal" (http://msbusiness.com/businessblog/2011/07/07/mississippi-cashes-in-on-bamboo/). *MS Business Journal.* . Retrieved 7 July 2011.

[10] Thomas R. Soderstrom; Cleofe E. Calderon; Thomas R. Soderstrom; Cleofe E. Calderon; T.R. Soderstrom, C.E. Calderon (1979). "A Commentary on the Bamboos (Poaceae: Bambusoideae)". *Biotropica* **11** (3): 161–172. doi:10.2307/2388036. JSTOR 2388036.

[11] Janzen, DH. (1976). "Why Bamboos Wait so Long to Flower". *Annual Review of Ecology and Systematics* **7**: 347–391. doi:10.1146/annurev.es.07.110176.002023.

[12] Keeley, JE; Keeley, J.E. and W.J. Bond (1999). "Mast flowering and semelparity in bamboos: The bamboo fire cycle hypothesis". *American Naturalist* **154** (3): 383–391. doi:10.1086/303243. PMID 10506551.

[13] Saha, S; Saha, S., HF Howe (2001). "The Bamboo Fire Cycle Hypothesis: A Comment". *The American Naturalist* **6** (158): 659–663. doi:10.1086/323593. PMID 18707360.

[14] Keeley, JE; Keeley, J.E. and W.J. Bond (2001). "On incorporating fire into our thinking about natural ecosystems: A response to Saha and Howe". *American Naturalist* **158** (6): 664–670. doi:10.1086/323594. PMID 18707361.

[15] http://www.britannica.com/EBchecked/topic/396788/muli-bamboo

[16] "Gorillas get drunk on bamboo sap" (http://www.telegraph.co.uk/earth/earthpicturegalleries/5037343/Gorillas-get-drunk-on-bamboo-sap.html). Telegraph.co.uk. 23 March 2009. . Retrieved 12 August 2009.

[17] Puri, H. S. (2003). *Rasayana ayurvedic herbs for longevity and rejuvenation* (http://books.google.com/?id=aQh25X9mzjAC&lpg=PP1&dq=Rasayana Ayurvedic Herbs for Longevity and Rejuvenation&pg=PP1#v=onepage&q=). London: Taylor & Francis. pp. 71–73. ISBN 0-203-21656-3. . Retrieved 12 August 2009.

[18] Peters, Tom F. (1987). *Transitions in Engineering: [[Guillaume Henri Dufour* (http://books.google.com/?id=73JPiTuDYscC)] and the Early 19th Century Cable Suspension Bridges]. Birkhauser. ISBN 3-7643-1929-1. .

[19] Landler, Mark (27 March 2002). "Hong Kong Journal; For Raising Skyscrapers, Bamboo Does Nicely" (http://www.nytimes.com/2002/03/27/world/hong-kong-journal-for-raising-skyscrapers-bamboo-does-nicely.html). New York Times. . Retrieved 12 August 2009.

[20] *Bamboo In Japan* (http://books.google.com/?id=UukQ2LaaP0wC&pg=PA101&lpg=PA101). Kodansha International. 1987. p. 101. ISBN 4-7700-2510-6. .

[21] Roger Lewis (1 July 2006). "Square Bamboo" (http://www.lewisbamboo.com/square.html). LewisBamboo.com. . Retrieved 12 August 2009.

[22] CASSANDRA ADAMS. "Bamboo Architecture and Construction with Oscar Hidalgo" (http://www.networkearth.org/naturalbuilding/bamboo.html). Natural Building Colloquium. . Retrieved 11 August 2009.

[23] Michelle Nijhuis (June 2009). "Bamboo Boom: Is This Material for You?" (http://www.scientificamerican.com/article.cfm?id=bamboo-boom). *Scientific American Earth 3.0 special.* Scientific American. . Retrieved 11 August 2009.

[24] Jonathan Bardelline (9 July 2009). "Growing the Future of Bamboo Products" (http://www.greenbiz.com/feature/2009/07/09/growing-future-bamboo-products). GreenBiz.com. . Retrieved 11 August 2009.

[25] "Bamboo Construction" (http://www.fastonline.org/CD3WD_40/VITA/BAMBOO/EN/BAMBOO.HTM). CD3WD. . Retrieved 11 August 2009.

[26] *Bamboo as a Building Material* (http://www.eric.ed.gov/ERICDocs/data/ericdocs2sql/content_storage_01/0000019b/80/34/73/06.pdf). Washington D.C.: US Department of Agriculture. 1981. pp. 7–11. . Retrieved 11 August 2009.

[27] http://www.greenschool.org

[28] Soni, Dr. K M (2011 [last update]). "India Pavilion at World Expo 2010" (http://www.nbmcw.com/reports/project-site-report/17869-india-pavilion-at-world-expo-2010.html). *NBM Media.* . Retrieved July 7, 2011.

[29] "Bamboo Fiber: Greenwash or Treasure?" (http://feelgoodstyle.com/2008/06/26/bamboo-fiber-greenwash-or-treasure/). Feelgood Style. 26 June 2008. . Retrieved 12 August 2009.

[30] "Competition Bureau Calls on Textile Dealers to Accurately Label Textile Articles Derived from Bamboo" (http://www.reuters.com/article/pressRelease/idUS175478+11-Mar-2009+MW20090311). Reuters. 11 March 2009. . Retrieved 12 August 2009.

[31] "Four Companies Charged with Labeling Rayon Clothing As Bamboo" (http://www.greenbiz.com/news/2009/08/11/companies-label-rayon-clothing-bamboo). GreenBiz.com. 11 August 2009. . Retrieved 12 August 2009.

[32] Perdue, Robert E.; Robert E. Perdue, Charles J. Kraebel, Tao Kiang (April 1961). "Bamboo Mechanical Pulp for Manufacture of Chinese Ceremonial Paper" (http://www.springerlink.com/content/9w17231j255j1178/). *Economic Botany* **15** (2): 161–164. doi:10.1007/BF02904089. . Retrieved 2009-08-14.

[33] Nanko, Hirko; Button, Allan; Hillman, Dave (2005). *The World of Market Pulp.* Appleton, WI, USA: WOMP, LLC. p. 256. ISBN 0-615-13013-5.

[34] Bhattacharya, S. (2010). *Tropical Bamboo: Molecular profiling and genetic diversity study.* Lambert Academic Publishing. ISBN 978-3-8383-7422-2.

[35] "Origins and development of bamboo music" (http://www.bbc.co.uk/learningzone/clips/origins-and-development-of-bamboo-music/11926.html). *bbc.co.uk.* 2011 [last update]. . Retrieved March 27, 2011.

[36] Jen Lukenbill. "About My Planet: Bamboo Bikes" (http://www.aboutmyplanet.com/environment/bamboo-bikes/). . Retrieved 2010-01-04.

[37] Bamboo: an untapped and amazing resource (http://www.unido.org/index.php?id=1000276) from UNIDO. Retrieved 30 November 2009.

[38] http://www.fao.org/documents/show_cdr.asp?url_file=/docrep/x5390e/x5390e05.htm

[39] Laws, B. 2010. Bamboo. Fifty Plants that Changed the Course of History.New York:Firefly Books (U.S)Inc.

[40] Chan, Alan Kam-leung and Gregory K. Clancey, Hui-Chieh Loy (2002). Historical Perspectives on East Asian Science, Technology and Medicine. Singapore: Singapore University Press. ISBN 9971-69-259-7. p. 15.

[41] Needham, Joseph (1986). *Science and Civilization in China*: Volume 3, *Mathematics and the Sciences of the Heavens and the Earth*. Taipei: Caves Books, Ltd. p. 614.

[42] Loewe, Michael (1997). "Wood and bamboo administrative documents of the Han period". In Edward L. Shaughnessy. *New Sources of Early Chinese History*. Society for the Study of Early China. pp. 161–192. ISBN 1-55729-058-X.

36. Soni, K M (2010), India Pavilion at World Expo 2010, Shanghai with Bamboo Dome. MGS Architecture, May–June issue, pp 59–63.

External links

- Chisholm, Hugh, ed. (1911). "Bamboo". *Encyclopædia Britannica* (11th ed.). Cambridge University Press.
- Bamboo Structural Design ISO Standards (http://www.iso.org/iso/catalogue_detail.htm?csnumber=36149)
- Resources on Bamboo (http://www.indiaenvironmentportal.org.in/category/thesaurus/climate-change/ environment/natural-disasters/ecosystems/forests/trees/bamboo)

Bambusoideae

Bamboos	
Bamboo forest in Kyoto, Japan	
Scientific classification	
Kingdom:	Plantae
Unranked:	Angiosperms
Unranked:	Monocots
Unranked:	Commelinids
Order:	Poales
Family:	Poaceae
Subfamily:	**Bambusoideae**
Genera	
See text.	

The **Bambusoideae** is a subfamily of the true grass family Poaceae, and is characterized by having 3 stigmas and are mostly tree-like.[1] However, there are uncertainties at practically every taxonomic level within the Bambusoideae, and different types of data (floral morphology, vegetative structures, anatomy, and genetics) often result in support for differing relationships.

The Bambusoideae generally consists of a distinct "core" group of genera, the woody bamboos (Bambuseae) and an associated group of genera of questionable affinity, the herbaceous bamboos (Bambusoideae). The bamusoid taxa have long been considered the most "primitive" grasses, mostly because of the presence of bracteate, indeterminate inflorescences, "pseudospikelets," and flowers with three lodicules, six stamens, and three stigmas.[2]

Taxonomy research

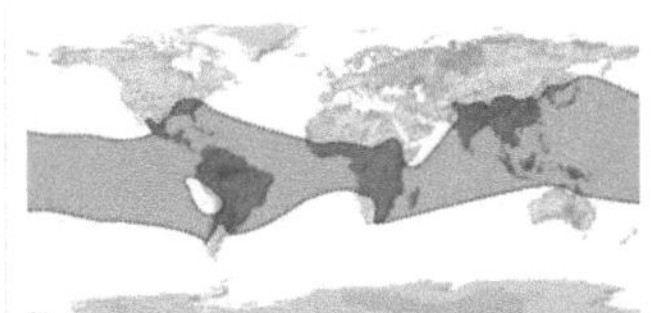

Flora distribution map of the Poaceae subfamily **Bambusoideae**.

In a recent study [3] DNA sequences for the chloroplast gene ndhF were analyzed to address phylogenetic relationships among the major lineages of the Poaceae. They found that two tribes of neotropical herbaceous bamboo tribes, the Streptochaeteae and Anomochloeae, are resolved as the most basal clade within the grass family, confirming the hypothesis that elements within the Bambusoideae sensu lato are basal within the Poaceae, and also showing that the Bambusoideae s.l. is polyphyletic.[2] A recent study which analyzed the phylogenetic relationships within the Bambusoideae using rp116 intron sequence data from chloroplast DNA was able to further resolve some of the uncertainties remaining in Clark et al.'s (1995) analysis. Kelchner and Clark's (1997) analysis resolved a Bambusoideae clade with two monophyletic groups: the Bambuseae (woody bamboos) and the Olyreae/Parianeae (herbaceous bamboos).[4]

Within the Bambuseae two clades were recovered corresponding to temperate and tropical woody bamboos, and the tropical taxa were even further divided into New World and Old World clades. The distinct lineages produced strongly correspond with geographic divisions, with major clades representing the New World herbaceous species (Olyreae/Parianeae), New World tropical woody bamboos, Old World tropical woody bamboos, and North temperate woody bamboos (all Bambuseae).

Subdivisions

Bambusoideae is divided into two groups:

1. the **Oryzodae**
2. the **Bambusodae**

Tribes

It has 13 tribes, as following:

- **Oryzodae**

   This group is also separated into subfamily Oryzaceae.

   - Tribe Anomochloeae

   This tribe is also separated to form subfamily Anomochlooideae. There is one genus: *Anomochloa*.
   - Tribe Diarrheneae

   There is one genus: *Diarrhena*.
   - Tribe Ehrharteae

   There are four genera: *Ehrharta, Microlaena, Petriella, Tetrarrhena*.
   - Tribe Olyreae

   There are 20 genera: *Agnesia, Arberella, Buergersiochloa, Cryptochloa, Diandrolyra, Ekmanochloa, Froesiochloa, Lithachne* (also placed in Oryzeae), *Maclurolyra, Mniochloa, Olyra (plant), Pariana, Parodiolyra, Piresia, Piresiella, Raddia, Raddiella, Rehia, Reitzia, Sucrea*.
   - Tribe Oryzeae

   This tribe also forms subfamily Oryzoideae. It has 13 genera: *Chikusichloa, Hydrochloa, Hygroryza, Leersia, Luziola, Maltebrunia, Oryza, Porteresia, Potamophila, Prosphytochloa, Rhynchoryza, Zizania, Zizaniopsis*.
   - Tribe Phaenospermatae

   There is 1 genus in this tribe: *Phaenosperma*.

- Tribe Phareae

  There are 4 genera: *Leptaspis, Pharus, Scrotochloa, Suddia.*
- Tribe Phyllorhachideae

  This tribe has 2 genera: *Humbertochloa, Phyllorhachis.*
- Tribe Streptochaeteae

  There is one genus: *Streptochaeta.*
- Tribe Streptogyneae

  This tribe only has one genus: *Streptogyna.*

- **Bambusodae**

  - Tribe Bambuseae

    This tribe comprises woody bamboos (or hardy bamboos). There are 91 genera, distributed into several subtribes:

    - Subtribe Arthrostylidiinae

    Comprises 13 genera: *Actinocladum, Alvimia, Apoclada, Arthrostylidium, Athroostachys, Atractantha, Aulonemia (Matudacalamus), Colanthelia, Elytrostachys, Glaziophyton, Merostachys, Myriocladus, Rhipidocladum.*

    - Subtribe Arundinariinae

    Comprises 16 genera: *Acidosasa, Ampelocalamus, Arundinaria, Borinda, Chimonocalamus, Drepanostachyum (Himalayacalamus), Fargesia, Ferrocalamus, Gaoligongshania, Gelidocalamus, Indocalamus, Oligostachyum, Pseudosasa, Sasa, Thamnocalamus, Yushania.*

    - Subtribe Bambusinae

    Comprises 10 genera: *Bambusa (Dendrocalamopsis), Bonia (Monocladus), Dendrocalamus (Klemachloa, Oreobambos, Oxynanthera* or *Sinocalamus), Dinochloa, Gigantochloa, Holttumochloa, Kinabaluchloa (Maclurochloa, Soejatmia), Melocalamus, Sphaerobambos, Thyrsostachys.*

    - Subtribe Chusqueinae

    Comprises 2 genera: *Chusquea, Neurolepis.*

    - Subtribe Guaduinae

    Comprises 4 genera: *Eremocaulon, Guadua, Olmeca, Otatea.*

    - Subtribe Melocanninae

    Comprises 9 genera: *Cephalostachyum, Davidsea, Leptocanna, Melocanna, Neohouzeaua, Ochlandra, Pseudostachyum, Schizostachyum, Teinostachyum.*

    - Subtribe Nastinae

    Comprises 6 genera: *Decaryochloa, Greslania, Hickelia, Hitchcockella, Nastus, Perrierbambus.*

    - Subtribe Racemobambodinae

    Comprises one genus: *Racemobambos (Neomicrocalamus, Vietnamosasa)*

    - Subtribe Shibataeinae

    Comprises 8 genera: *Chimonobambusa, Indosasa, Phyllostachys, Qiongzhuea, Semiarundianria (Brachystachyum), Shibataea, Sinobambusa, Temburongia* (incertae sedis).

  - Tribe Guaduelleae

    There is one genus: *Guaduella.*
  - Tribe Puelieae

    There is one genus: *Puelia.*

References

[1] Judd, WS, CS Campbell, EA Kellogg, PF Stevens, MJ Donoghue [eds.]. 2008. Plant Systematics: A Phylogenetic Approach, 296-301. Sinauer Associates, Inc., Sunderland, Massachusetts USA.

[2] Clark, LG, W Zhang, JF Wendel. 1995. A Phylogeny of the Grass Family (Poaceae) Based on ndhF Sequence Data. Systematic Botany 20(4): 436-460.

[3] (Clark et al., 1995)

[4] Kelchner, SA, LG Clark. 1997. Molecular Evolution and Phylogenetic Utility of the Chloroplast rpl16 Intron in Chusquea and the Bambusoideae (Poaceae). Molecular Phylogenetics and Evolution 8(3): 385-397.

Poaceae

<table>
<tr><td colspan="2" align="center">Poaceae (true grasses)
Temporal range: Late Cretaceous[1] - Recent,</td></tr>
<tr><td colspan="2"></td></tr>
<tr><td colspan="2" align="center">Flowering head of Meadow Foxtail (Alopecurus pratensis), with stamens exerted at anthesis</td></tr>
<tr><td colspan="2" align="center">Scientific classification</td></tr>
<tr><td>Kingdom:</td><td>Plantae</td></tr>
<tr><td>Unranked:</td><td>Angiosperms</td></tr>
<tr><td>Unranked:</td><td>Monocots</td></tr>
<tr><td>Unranked:</td><td>Commelinids</td></tr>
<tr><td>Order:</td><td>Poales</td></tr>
<tr><td>Family:</td><td>Poaceae
Barnhart</td></tr>
<tr><td colspan="2" align="center">Type genus</td></tr>
<tr><td colspan="2" align="center">Poa
L.</td></tr>
<tr><td colspan="2" align="center">Subfamilies</td></tr>
<tr><td colspan="2">There are 12 (or 13) subfamilies:
Subfamily Anomochlooideae
Subfamily Pharoideae
Subfamily Puelioideae
Subfamily Bambusoideae
Subfamily Pooideae
Subfamily Ehrhartoideae
Subfamily Aristidoideae
Subfamily Arundinoideae
Subfamily Centothecoideae, sometimes included in Panicoideae
Subfamily Chloridoideae
Subfamily Panicoideae
Subfamily Danthonioideae
Subfamily Micrairoideae</td></tr>
</table>

The **Poaceae** (also known as the **Gramineae**) is a large and nearly ubiquitous family of monocot flowering plants. Members of this family are commonly called (land) **grasses**, although the term (land) "grass" is also applied to plants that are not in the Poaceae lineage, including the rushes (Juncaceae) and sedges (Cyperaceae). As for the **seagrasses**, they all belong to the Alismatales, a different monocot order altogether. This broad and general use of the word "grass" has led to plants of the Poaceae often being called "true grasses". With over 10,025 currently accepted species, the Poaceae represent the fifth largest plant family. Only the Orchidaceae, Asteraceae, Fabaceae, and Rubiaceae have more species.[2]

Plant communities dominated by Poaceae are called grasslands; grasslands are estimated to comprise 20% of the vegetation cover of the Earth. Grass species also occur in many other habitats not formally considered to be grasslands, including different types of wetlands (e.g., fens, marshes), forests and tundra.

Poaceae are often considered to be the most important of all plant families to human economies: it includes the staple food grains and cereal crops grown around the world, lawn and forage grasses, and bamboo, which is widely used for construction throughout east Asia and sub-Saharan Africa. Civilization was founded largely on the ability to domesticate cereal grass crops around the world.

Description

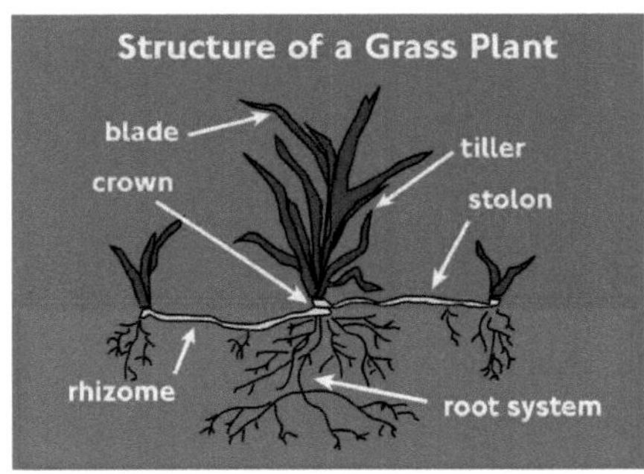

Structure of a grass plant

Grasses generally have the following characteristics (the image gallery can be used for reference):

Poaceae have hollow stems called culms, which are plugged (solid) at intervals called nodes, the points along the culm at which leaves arise. Grass leaves are alternate, distichous (in one plane) or rarely spiral, and parallel-veined. Each leaf is differentiated into a lower sheath, which hugs the stem for a distance and a blade with margins usually entire. The leaf blades of many grasses are hardened with silica phytoliths, which helps discourage grazing animals. In some grasses (such as sword grass), this makes the edges of the grass blades sharp enough to cut human skin. A membranous appendage or fringe of hairs, called the ligule, lies at the junction between sheath and blade, preventing water or insects from penetrating into the sheath.

Flowers of Poaceae are characteristically arranged in spikelets, each spikelet having one or more florets (the spikelets are further grouped into panicles or spikes). A spikelet consists of two (or sometimes fewer) bracts at the base, called glumes, followed by one or more florets. A floret consists of the flower surrounded by two bracts called the lemma (the external one) and the palea (the internal). The flowers are usually hermaphroditic (maize, monoecious, is an exception) and pollination is always

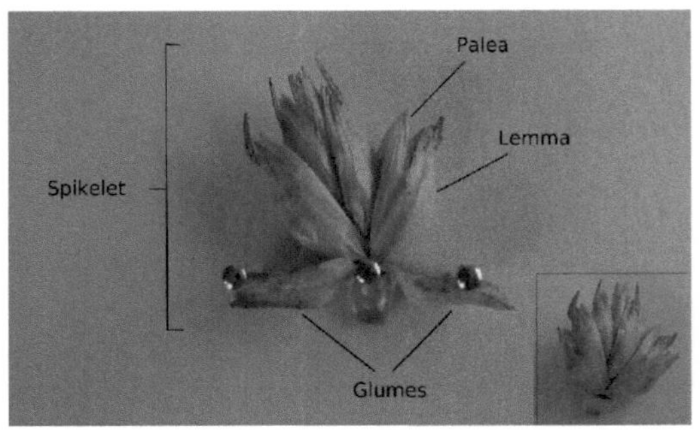

Parts of a spikelet

anemophilous, that is, by wind. The perianth is reduced to two scales, called lodicules, that expand and contract to spread the lemma and palea; these are generally interpreted to be modified sepals. This complex structure can be seen in the image on the right, portraying a wheat (*Triticum aestivum*) spike.

The fruit of Poaceae is a caryopsis, in which the seed coat is fused to the fruit wall and thus, not separable from it (as in a maize kernel).

A tiller a non-seed leaf shoot.

Growth and development

Grass blades grow at the base of the blade and not from elongated stem tips. This low growth point evolved in response to grazing animals and allows grasses to be grazed or mown regularly without severe damage to the plant.[3]

Three general classifications of growth habit present in grasses: bunch-type (also called caespitose), stoloniferous, and rhizomatous.

The success of the grasses lies in part in their morphology and growth processes, and in part in their physiological diversity. Most of the grasses divide into two physiological groups, using the C3 and C4 photosynthetic pathways for carbon fixation. The C4 grasses have a photosynthetic pathway linked to

Grass flowers

specialized Kranz leaf anatomy that particularly adapts them to hot climates and an atmosphere low in carbon dioxide.

C3 grasses are referred to as "cool season" grasses, while C4 plants are considered "warm season" grasses; they may be either annual or perennial.

- Annual cool season - wheat, rye, annual bluegrass (annual meadowgrass, *Poa annua*), and oat
- Perennial cool season - orchardgrass (cocksfoot, *Dactylis glomerata*), fescue (*Festuca* spp), Kentucky bluegrass and perennial ryegrass (*Lolium perenne*)
- Annual warm season - corn, sudangrass, and pearl millet
- Perennial warm season - big bluestem, indiangrass, bermudagrass and switchgrass.[4]

Ecology

Biomes dominated by grasses are called grasslands. If only large contiguous areas of grasslands are counted, these biomes cover 31% of the planet's land.[5] Grasslands go by various names depending on location, including pampas, plains, steppes, or prairie.

In addition to their use as forage worldwide by many grazing mammals, such as cattle and other livestock, deer, and elephants, grasses are used as food plants by many species of butterflies and moths; see List of Lepidoptera that feed on grasses.

The evolution of large grazing animals in the Cenozoic has contributed to the spread of grasses. Without large grazers, a clearcut of fire-destroyed area would soon be colonized by grasses and, if there is enough rain, tree seedlings. The tree seedlings would eventually produce shade, which kills most grasses. Large animals, however, trample the seedlings, killing the trees. Grasses persist because their lack of woody stems helps them to resist the damage of trampling.[6]

Evolution

Until recently, grasses were thought to have evolved around 55 million years ago, based on fossil records. However, recent findings of 65-million-year-old phytoliths resembling grass phytoliths (including ancestors of rice and bamboo) in Cretaceous dinosaur coprolites,[1] [7] may place the diversification of grasses to an earlier date. Indeed, revised dating of the origins of the rice tribe Oryzeae have been led to the suggestion that the date might be pushed back as early as 107 Ma to 129 Ma.[8]

The relationships among the subfamilies *Bambusoideae*, *Ehrhartoideae* and *Pooideae* in the BEP clade have been resolved: Bambusoideae and Pooideae are more closely related than Ehrhartoideae.[9] This separation occurred within a relatively short time span (~4 million years).

Distribution

The grass family is one of the most widely distributed and abundant groups of plants on Earth. They are found on every continent, and are essentially only absent from central Greenland and much of Antarctica.[2]

Taxonomy

The most recent classification of the grass family *A World-wide Phylogenetic Classification of Poaceae* [10]|date=March 2012, recognizes 12 subfamilies:

Setaria verticillata from Panicoideae

- Anomochlooideae, a small lineage of broad-leaved grasses that includes two genera (*Anomochloa*, *Streptochaeta*)
- Pharoideae, a small lineage of grasses that includes three genera, including *Pharus* and *Leptaspis*
- Puelioideae, a small lineage that includes the African genus *Puelia*
- Pooideae, including wheat, barley, oats, brome-grass (*Bromus*), reed-grasses (*Calamagrostis*) and many lawn and pasture grasses
- Bambusoideae, including bamboo
- Ehrhartoideae, including rice, wild rice
- Aristidoideae, including [*Aristida*]
- Arundinoideae, including giant reed, common reed
- Chloridoideae, including the lovegrasses (*Eragrostis*, about 350 species, including teff), dropseeds (*Sporobolus*, some 160 species), finger millet (*Eleusine coracana* (L.) Gaertn.), and the muhly grasses (*Muhlenbergia*, about 175 species)
- Panicoideae, including Centothecoideae, panic grass, maize, sorghum, sugarcane, most millets, fonio, and bluestem grasses
- Micrairoideae
- Danthonioideae, including pampas grass

Depending on the classification followed, the family includes approximately 668 genera.[2]

Tragus roxburghii from Chloridoideae

Etymology

The Poaceae was named by John Hendley Barnhart in 1895,[11] based on the tribe Poeae (described in 1814 by Robert Brown), and the type genus *Poa* (described in 1753 by Linnaeus). The term is derived from the Ancient Greek term for "grass".

Uses

Grasses are, in human terms, perhaps the most economically important plant family. Grasses' economic importance stems from several areas, including food production, industry, and lawns.

Food production

Agricultural grasses grown for their edible seeds are called cereals. Three cereals – rice, wheat, and maize (corn) – provide more than half of all calories eaten by humans.[12] Of all crops, 70% are grasses.[13] Cereals constitute the major source of carbohydrates for humans and perhaps the major source of protein, and include rice in southern and eastern Asia, maize in Central and South America, and wheat and barley in Europe, northern Asia and the Americas.

Sugarcane is the major source of sugar production. Many other grasses are grown for forage and fodder for animal feed, particularly for sheep and cattle, thereby indirectly providing more human calories.

Industry

Grasses are used for construction. Scaffolding made from bamboo is able to withstand typhoon-force winds that would break steel scaffolding.[5] Larger bamboos and *Arundo donax* have stout culms that can be used in a manner similar to timber, and grass roots stabilize the sod of sod houses. *Arundo* is used to make reeds for woodwind instruments, and bamboo is used for innumerable implements.

Grass fiber can be used for making paper, and for biofuel production.

Phragmites australis (common reed) is important in water treatment, wetland habitat preservation and land reclamation in Afro-Eurasia.

Lawn and ornamental grasses

Grasses are the primary plant used in lawns, which themselves derive from grazed grasslands in Europe. They also provide an important means of erosion control (e.g., along roadsides), especially on sloping land.

Although supplanted by artificial turf in some games, grasses are still an important covering of playing surfaces in many sports, including football, tennis, golf, cricket, and softball/baseball.

Ornamental grasses, such as perennial bunch grasses, are used in many styles of garden design for their foliage, inflorescences, seed heads, and slope stabilization. They are often used in natural landscaping, xeriscaping, contemporary or modern landscaping, wildlife gardening, and native plant gardening.

Economically important grasses

Grain crops	Leaf and stem crops	Lawn grasses	Ornamental grasses (Horticultural)	Model organisms
• Barley	• Bamboo	• Bahia grass	• Calamagrostis spp.	• *Brachypodium distachyon*
• Maize (Corn)	• Marram grass	• Bent grass	• Cortaderia spp.	• Maize (Corn)
• Oats	• Meadow-grass	• Bermuda grass	• Deschampsia spp.	• Rice
• Rice	• Reed	• Centipede grass	• Festuca spp.	• Sorghum
• Rye	• Ryegrass	• Fescue	• Melica spp.	• Wheat
• Sorghum	• Sugarcane	• Meadow-grass	• Muhlenbergia spp.	
• Wheat		• Ryegrass	• Stipa spp.	
• Millet		• St. Augustine grass		
		• Zoysia		

Grasses and society

Grasses have long had significance in human society. They have been cultivated as a food source for domesticated animals for up to 10,000 years, and have been used to make paper since the second century AD. Also, the primary ingredient of beer is usually barley or wheat, both of which have been used for this purpose for over 4,000 years.

Grass-covered house in Iceland

Some common aphorisms involve grass. For example:

- "The grass is always greener on the other side" suggests an alternate state of affairs will always seem preferable to one's own.
- "Don't let the grass grow under your feet" tells someone to get moving.
- "A snake in the grass" means dangers that are hidden.
- "When elephants fight, it is the grass which suffers" tells of bystanders caught in the crossfire.

A folk myth about grass is that it refuses to grow where any violent death has occurred.[14]

Genera

See the full List of Poaceae genera.

Image gallery

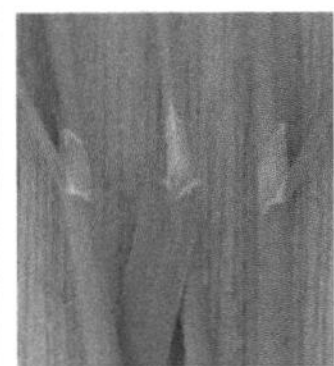
Leaves of *Poa trivialis* showing the ligules

Bamboo stem and leaves, nodes are evident

A *Chasmanthium latifolium* spikelet

Wheat spike and spikelet

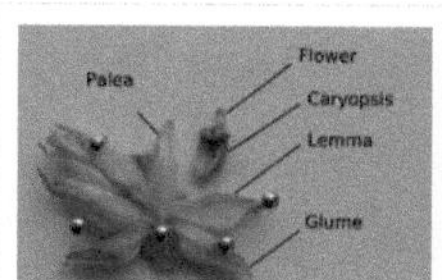

Spikelet opened to show caryopsis

Harestail grass

Grass

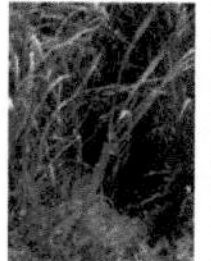

Sugarcane (*Saccharum officinarum*)

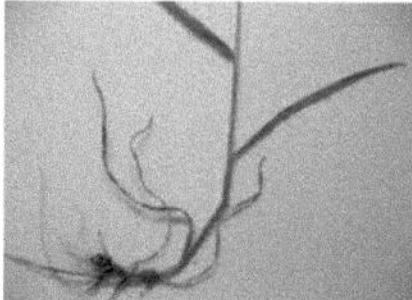

Roots of *Bromus hordeaceus*

Barley mature spikes (*Hordeum vulgare*)

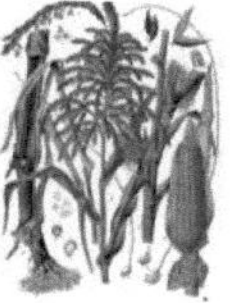

Illustration depicting both staminate and pistillate flowers of maize (*Zea mays*)

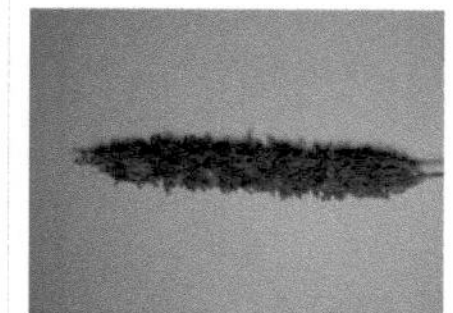

A grass flower head (meadow foxtail) showing the plain-coloured flowers with large anthers.

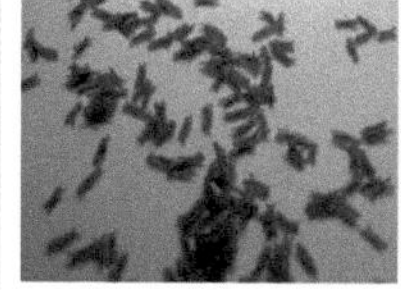

Anthers detached from a meadow foxtail flower

Setaria verticillata, bristly foxtail

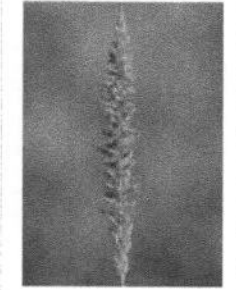

Setaria verticillata, bristly foxtail

See also

- Agrostology
- Grass
- Bunch grass
- Ornamental grass
- Sedges
- Rushes

References

[1] Piperno, D. R.; Sues, H.D. (2005). "Dinosaurs Dined on Grass". *Science* **310** (5751): 1126hor = Piperno, D.R.. doi:10.1126/science.1121020. PMID 16293745.

[2] Stevens, P.F. "Angiosperm Phylogeny Website" (http://www.mobot.org/MOBOT/Research/APweb/welcome.html#Famlarge). . Retrieved 2007-10-07.

[3] David Attenborough (1984). *The Living Planet*. British Broadcasting Corporation. pp. 113–4. ISBN 0-563-20207-6.

[4] (http://forages.oregonstate.edu/projects/regrowth/main.cfm?PageID=33)

[5] George Constable (ed), ed. (1985). *Grasslands and Tundra*. Planet Earth. Time Life Books. p. 20. ISBN 0-8094-4520-4.

[6] David Attenborough (1984). *The Living Planet*. British Broadcasting Corporation. p. 137.

[7] Prasad, V.; Stroemberg, C.A.E.; Alimohammadian, H.; Sahni, A. (2005). "Dinosaur Coprolites and the Early Evolution of Grasses and Grazers". *Science(Washington)* **310** (5751): 1177–1180. doi:10.1126/science.1118806. PMID 16293759.

[8] Prasad V, Strömberg CA, Leaché AD, Samant B, Patnaik R, Tang L, Mohabey DM, Ge S, Sahni A. (2011). Late Cretaceous origin of the rice tribe provides evidence for early diversification in Poaceae. Nat Commun. 2:480. doi:10.1038/ncomms1482 PMID 21934664

[9] Wu ZQ, Ge S (2011) The phylogeny of the BEP clade in grasses revisited: Evidence from the whole-genome sequences of chloroplasts. Mol Phylogenet Evol

[10] http://www.tropicos.org/projectwebportal.aspx?pagename=ClassificationNWG&projectid=10

[11] Barnhart, J.H. (1895) Poaceae. *Bulletin of the Torrey Botanical Club* **22**: 7.

[12] Peter H. Raven & George B. Johnson (1995). Carol J. Mills (ed). ed. *Understanding Biology* (3rd ed.). WM C. Brown. p. 536. ISBN 0-697-22213-6.

[13] George Constable (ed), ed. (1985). *Grasslands and Tundra*. Planet Earth. Time Life Books. p. 19. ISBN 0-8094-4520-4.

[14] Olmert, Michael (1996). *Milton's Teeth and Ovid's Umbrella: Curiouser & Curiouser Adventures in History*, p. 208. Simon & Schuster, New York. ISBN 0-684-80164-7.

External links

- Poaceae (http://www.theplantlist.org/browse/A/Poaceae/) at *The Plant List* (http://www.theplantlist.org/)
- Gramineae (http://delta-intkey.com/angio/www/graminea.htm) at *The Families of Flowering Plants (DELTA)* (http://delta-intkey.com/angio/)
- Poaceae (http://eol.org/pages/8223/overview) at the *Encyclopedia of Life* (http://eol.org/)
- Poaceae (http://www.mobot.org/mobot/research/apweb/orders/poalesweb.htm#Poaceae) at the *Angiosperm Phylogeny Website* (http://www.mobot.org/mobot/research/apweb/)
- *Poaceae Classification* (http://www.tropicos.org/projectwebportal.aspx?pagename=ClassificationNWG& projectid=10) from the online *Catalogue of New World Grasses* (http://www.tropicos.org/Project/CNWG)
- Poaceae (http://www.efloras.org/florataxon.aspx?flora_id=2&taxon_id=10711) at the online *Flora of China* (http://www.efloras.org/flora_page.aspx?flora_id=2)
- Poaceae (http://greif.uni-greifswald.de/floragreif/?flora_search=Taxon&action=genus&fam=Poaceae) at the online *Guide to the Flora of Mongolia* (http://greif.uni-greifswald.de/floragreif/)
- Poaceae (http://www.efloras.org/flora_page.aspx?flora_id=1050) at the online *Flora of Taiwan* (http://www. efloras.org/flora_page.aspx?flora_id=100)
- Poaceae (http://www.efloras.org/florataxon.aspx?flora_id=5&taxon_id=10711) at the online *Flora of Pakistan* (http://www.efloras.org/flora_page.aspx?flora_id=5)
- Poaceae (http://www.zimbabweflora.co.zw/speciesdata/family.php?family_id=177) at the online *Flora of Zimbabwe* (http://www.zimbabweflora.co.zw/index.php)
- Poaceae (http://florabase.dec.wa.gov.au/browse/profile/22751) at the online *Flora of Western Australia* (http://florabase.dec.wa.gov.au/)
- Grasses of Australia (AusGrass2) - http://ausgrass2.myspecies.info/
- Gramineae (http://floraseries.landcareresearch.co.nz/pages/Taxa. aspx?id=_8ae81d6d-1dca-418b-8bcf-4b6e1e5a2f6d&fileName=Flora 5.xml) at the online *Flora of New Zealand* (http://floraseries.landcareresearch.co.nz/pages/index.aspx)
- NZ Grass Key (http://www.landcareresearch.co.nz/research/biosystematics/plants/grasskey/) An Interactive Key to New Zealand Grasses at *Landcare Research* (http://www.landcareresearch.co.nz/)

- The Grass Genera of the World (http://delta-intkey.com/grass/) at *DELTA intkey* (http://delta-intkey.com/)
- GrassBase - The Online World Grass Flora (http://www.rbgkew.org.uk/data/grasses-db/sppindex.htm) at the *Royal Botanic Gardens - Kew* (http://www.kew.org/)
- GrassWorld - http://grassworld.myspecies.info/

Poales

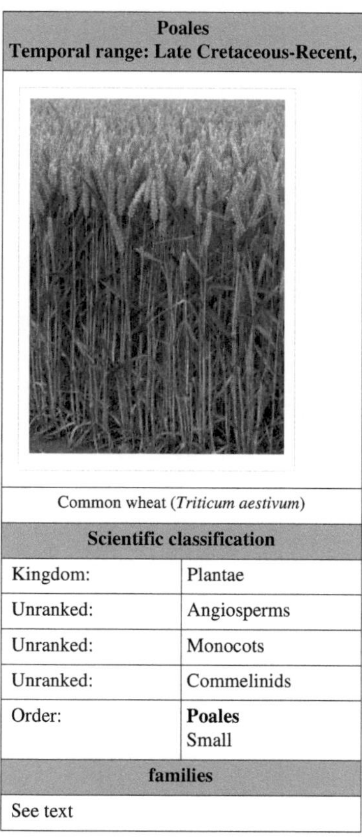

Common wheat (*Triticum aestivum*)

Poales Temporal range: Late Cretaceous-Recent,	
Scientific classification	
Kingdom:	Plantae
Unranked:	Angiosperms
Unranked:	Monocots
Unranked:	Commelinids
Order:	**Poales** Small
families	
See text	

Poales is a large order of flowering plants in the monocotyledons, and includes families of plants such as the grasses, bromeliads, and sedges. Sixteen plant families are currently recognized by botanists to be part of Poales.

The earliest fossils attributed to the Polaes dates to the late Cretaceous period about 65.5 million years ago, though some studies (e.g., Bremer, 2002) suggest the origin of the of the group may extend to nearly 115 million years ago, likely in South America. The earliest known fossils include pollen and fruits. The flowers are typically small, enclosed by bracts, and arranged in an inflorescence (except in the genus Mayaca, with solitary terminal flowers). The flowers of many species are wind pollinated; the seeds usually contain starch.

Billbergia pyramidalis of family Bromeliaceae

The APG III system (2009) accepts the order and places it in a clade called commelinids, in the monocots. It uses this circumscription:

- order Poales
    - family Anarthriaceae
    - family Bromeliaceae
    - family Centrolepidaceae
    - family Cyperaceae
    - family Ecdeiocoleaceae
    - family Eriocaulaceae

Setaria verticillata from family Poaceae

    - family Flagellariaceae
    - family Joinvilleaceae
    - family Juncaceae
    - family Mayacaceae
    - family Poaceae
    - family Rapateaceae
    - family Restionaceae
    - family Thurniaceae
    - family Typhaceae
    - family Xyridaceae

The earlier APG system (1998) adopted the same placement, although it used the spelling "commelinoids", and used the following circumscription (i.e., it did not include the plants in families Bromeliaceae and Mayacaceae in the order):

- order Poales
    - family Anarthriaceae
    - family Centrolepidaceae
    - family Cyperaceae
    - family Ecdeiocoleaceae
    - family Eriocaulaceae

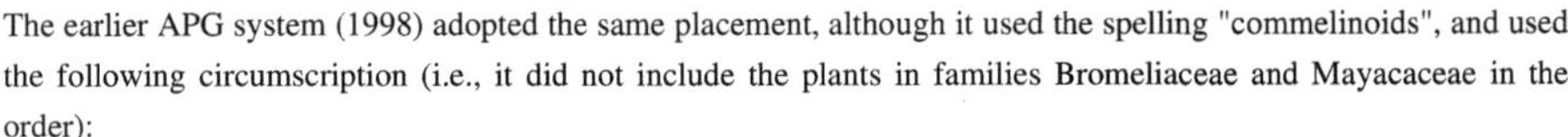

family Flagellariaceae

family Hydatellaceae (now transferred out of the monocots; recently discovered to be an 'early-diverging' lineage of flowering plants.)

family Joinvilleaceae

family Juncaceae

family Poaceae

family Prioniaceae

family Restionaceae

family Sparganiaceae (now included in family Typhaceae.)

family Thurniaceae

family Typhaceae

family Xyridaceae

The morphology-based Cronquist system did not include an order named Poales, assigning these families to the orders Bromeliales, Cyperales, Hydatellales, Juncales, Restionales and Typhales.

In early systems an order including the grass family did not go by the name Poales but by a descriptive botanical name such as Graminales in the Engler system (update of 1964) and in the Hutchinson system (first edition, first volume, 1926), Glumiflorae in the Wettstein system (last revised 1935) or Glumaceae in the Bentham & Hooker system (third volume, 1883).

Uses

The Poales is the most economically important order of monocots and possibly the most important order of plants in general. Within the order, by far the most important family economically is the family of grasses (Poaceae, syn. Gramineae), which includes barley, maize, millet, rice, and wheat. It is also the largest family in the order, far outnumbering its competitors:

- Poaceae: 12,070 species
- Cyperaceae: 5,500 species
- Bromeliaceae: 3,170 species
- Eriocaulaceae: 1,150 species

References

- Bremer, K. (2002). Gondwanan Evolution of the Grass Alliance of Families (Poales). *Evolution* 56: 1374-1387. [Available online: Abstract [1]]
- Judd, W. S., C. S. Campbell, E. A. Kellogg, P. F. Stevens, M. J. Donoghue (2002). *Plant Systematics: A Phylogenetic Approach, 2nd edition.* pp. 276-292 (Poales). Sinauer Associates, Sunderland, Massachusetts. ISBN 0-87893-403-0 .
- Linder, H. Peter and Paula J. Rudall. 2005. Evolutionary History of the Poales. *Annual Review of Ecology, Evolution, and Systematics* 36: 107-124.
- Small, J. K. (1903). *Flora of the Southeastern United States*, 48. New York, U.S.A.

External links

- NCBI Taxonomy Browser [2]
- APWeb [3]

References

[1] http://www.bioone.org/bioone/?request=get-abstract&issn=0014-3820&volume=056&issue=07&page=1374

[2] http://www.ncbi.nlm.nih.gov/Taxonomy/Browser/wwwtax.cgi?mode=Tree&id=38820&lvl=3&lin=f&keep=1&srchmode=1&
     unlock

[3] http://www.mobot.org/MOBOT/Research/APweb/orders/Poalesweb.htm#Poales

Commelinids

<table>
<tr><td colspan="2" align="center">commelinids</td></tr>
<tr><td colspan="2" align="center"></td></tr>
<tr><td colspan="2" align="center">Scientific classification</td></tr>
<tr><td>Kingdom:</td><td>Plantae</td></tr>
<tr><td>Unranked:</td><td>Angiosperms</td></tr>
<tr><td>Unranked:</td><td>Monocots</td></tr>
<tr><td>Unranked:</td><td>Commelinids</td></tr>
<tr><td colspan="2" align="center">Orders</td></tr>
<tr><td colspan="2">
<ul>
<li>Arecales</li>
<li>Commelinales</li>
<li>Poales</li>
<li>Zingiberales</li>
<li>unplaced family Dasypogonaceae</li>
</ul>
</td></tr>
</table>

In plant taxonomy, **commelinids** (plural, not capitalised) is a name used by the APG III system for a clade within the monocots, which in its turn is a clade within the angiosperms. The commelinids are the only clade that the APG has informally named within the monocots. The remaining monocots are a paraphyletic unit.

Members of the commelinid clade have cell walls containing UV-fluorescent ferulic acid.[1]

Classification

The commelinids were first recognized as a formal group in 1967 by Armen Takhtajan, who named them the Commelinidae and assigned them to a subclass of the monocots.[2] However, by the release of his 1980 system of classification, he had merged this subclass into a larger one no longer considered to be a clade.

The commelinids constitute a well-supported as a clade within the monocots,[3] and this clade has been recognized in all three APG classification systems. The commelinids of APG II (2003) and APG III (2009) contain essentially the same plants as the commelinoids of the earlier APG system (1998).[3]

clade monocots :

- clade commelinids:
    - family Dasypogonaceae
- order Arecales (palms)
- order Commelinales (spiderwort, water hyacinth)
- order Poales (grasses, rushes, bromeliads)
- order Zingiberales (gingers, banana)

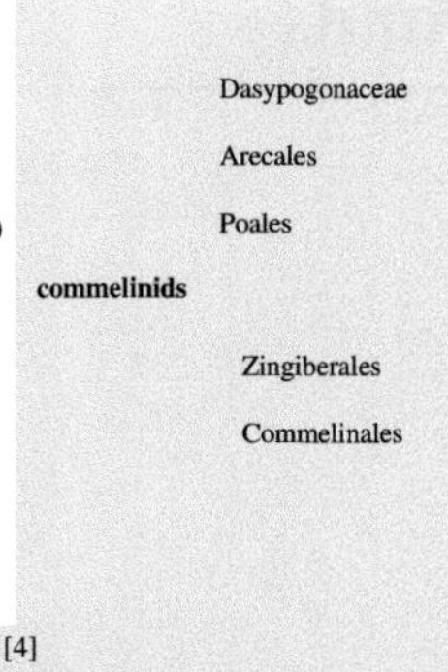

The current phylogeny and composition of the commelinids.[4]

References

[1] Dahlgren, R. M. T.; Rassmussen, F. (1983). "Monocotyledon evolution. Characters and phylolgenetic estimation". *Evol. Biol.* **16**: 255–395.

[2] Takhtajan, A. (1967). *Система и филогения цветкорых растений (Systema et Phylogenia Magnoliophytorum)*. Moscow: Nauka.

[3] Cantino, Philip D.; James A. Doyle, Sean W. Graham, Walter S. Judd, Richard G. Olmstead, Douglas E. Soltis, Pamela S. Soltis, & Michael J. Donoghue (2007). "Towards a phylogenetic nomenclature of *Tracheophyta*". *Taxon* **56** (3): E1–E44.

[4] "An update of the Angiosperm Phylogeny Group classification for the orders and families of flowering plants: APG III". *Botanical Journal of the Linnean Society* **161**: 105–121. 2009. doi:10.1111/j.1095-8339.2009.00996.x.

Monocotyledon

Monocotyledons	
Hemerocallis flower, with three flower parts in each whorl	
Scientific classification	
Kingdom:	Plantae
Unranked:	Angiosperms
Unranked:	**Monocots**
Orders	
About 10; *see Classification*	

Monocotyledons, also known as **monocots**, are one of two major groups of flowering plants (or angiosperms) that are traditionally recognized, the other being dicotyledons, or dicots. Monocot seedlings typically have one cotyledon (seed-leaf), in contrast to the two cotyledons typical of dicots. Monocots have been recognized at various taxonomic ranks, and under various names (see below). The APG II system recognises a clade called "monocots" but does not assign it to a taxonomic rank.

Wheat, an economically important monocot

According to the IUCN there are 59,300 species of monocots.[1] The largest family in this group (and in the flowering plants as a whole) by number of species are the orchids (family Orchidaceae), with more than 20,000 species.[2] In agriculture the majority of the biomass produced comes from monocots.[3] The true grasses, family Poaceae (Gramineae), are the most economically important family in this group. These include all the true grains (rice, wheat, maize, etc.), the pasture grasses, sugar cane, and the bamboos. True grasses have evolved to become highly specialised for wind pollination. Grasses produce much smaller flowers, which are gathered in highly visible plumes (inflorescences). Other economically important monocot families are the palm family (Arecaceae), banana family (Musaceae), ginger family (Zingiberaceae) and the onion family Alliaceae, which includes such ubiquitously used vegetables as onions and garlic.

Many plants cultivated for their blooms are also from the monocot group, notably lilies, daffodils, irises, amaryllis, orchids, cannas, bluebells and tulips.

Name, characters

The name monocotyledons is derived from the traditional botanical name "Monocotyledones", which derives from the fact that most members of this group have one cotyledon, or embryonic leaf, in their seeds. By contrast, the traditional dicotyledons typically have two cotyledons. From a diagnostic point of view the number of cotyledons is neither a particularly handy (as they are only present for a very short period in a plant's life), nor totally reliable character.

Nevertheless, monocots are a distinctive group.[4] One of the most noticeable traits is that a monocot's flower is trimerous, with the flower parts in threes or in multiples of three. That is to say, a monocotyledon's flower typically has three, six, or nine petals. Many monocots also have leaves with parallel veins.

Morphology, compared to the (broadly defined) dicotyledons

The traditionally listed differences between monocotyledons and dicotyledons are as follows. This is a broad sketch only, not invariably applicable, as there are a number of exceptions. The differences indicated are more true for monocots versus eudicots.[4]

Hypoxis decumbens L. with a typical monocot perigone and parallel leaf venation

Slice of onion, showing parallel veins in cross section

Ceroxylon quindiuense (Quindio wax palm) is considered the tallest monocot in the world

Feature	In monocots	In dicots
Number of parts of each flower	in threes (flowers are trimerous)	in fours or fives (tetramerous or pentamerous)
Number of furrows or pores in pollen	one	three
Number of cotyledons (leaves in the seed)	one	two
Arrangement of vascular bundles in the stem	scattered	in concentric circles
Roots	are adventitious	develop from the radicle

Arrangement of major leaf veins	parallel	reticulate

The vast majority of Monocots lack a petiole in their leaves.

A number of these differences are not unique to the monocots. For example, trimerous flowers and monosulcate pollen are also found in magnoliids.[4] Exclusively adventitious roots are found also in Nymphaeaceae and some of the Piperaceae.[4] Similarly, at least one of these traits, parallel leaf veins, is far from universal among the monocots. Monocots with reticulate leaf veins are found in a wide variety of monocot families: for example, *Trillium*, *Smilax* (greenbriar), and *Pogonia* (an orchid), and the Dioscoreales.[4] Nevertheless, this list of traits is a generally valid set of contrasts, especially when contrasting monocots with eudicots rather than non-monocot flowering plants in general.[4]

Emergence

Some monocots, such as grasses, have hypogeal emergence, where the mesocotyl elongates and pushes the coleoptile (which encloses and protects the shoot tip) toward the soil surface.[5] Since elongation occurs above the cotyledon, it is left in place in the soil where it was planted. Many dicots have epigeal emergence, in which the hypocotyl elongates and becomes arched in the soil. As the hypocotyl continues to elongate, it pulls the cotyledons upward, above the soil surface.

Vascular system

Monocots have a distinctive arrangement of vascular tissue known as an atactostele in which the vascular tissue is scattered rather than arranged in concentric rings. Many monocots are herbaceous and do not have the ability to increase the width of a stem (secondary growth) via the same kind of vascular cambium found in non-monocot woody plants.[4] However, some monocots do have secondary growth, and because it does not arise from a single vascular cambium producing xylem inwards and phloem outwards, it is termed "anomalous secondary growth".[6] Examples of large monocots which either exhibit secondary growth, or can reach large sizes without it, are palms (Arecaceae), screwpines (Pandanaceae), bananas (Musaceae), *Yucca*, *Aloe*, *Dracaena*, and *Cordyline*.[4]

Stems of two *Roystonea regia* palms showing anomalous secondary growth in monocots. Note the characteristic fibrous roots, typical of monocots.

Classification

The monocots are considered to form a monophyletic group arising early in the history of the flowering plants. The earliest fossils presumed to be monocot remains date from the early Cretaceous period.

Taxonomists have considerable latitude in naming this group, as the monocots are a group above the rank of family. Article 16 of the *ICBN* allows either a descriptive name or a name formed from the name of an included family.

Historically, the monocotyledons were named:

- Monocotyledoneae in the de Candolle system and the Engler system,
- Monocotyledones in the Bentham & Hooker system and the Wettstein system,
- class Liliopsida in the Takhtajan system and the Cronquist system,
- subclass Liliidae in the Dahlgren system and the Thorne system (1992), and
- clade monocots in the APG system and the APG II system.

Each of these systems uses its own internal taxonomy for the group. The monocotyledons are famous as a group that is extremely stable in its outer borders (it is a well-defined, coherent group), while in its internal taxonomy is extremely unstable (historically no two authoritative systems have agreed with each other on how the monocotyledons are related to each other).

Recent molecular studies have both confirmed the monophyly of the monocots and helped elucidate relationships within this group. The APG II system does not assign the monocots to a taxonomic rank, instead recognizing a monocots clade. This system recognizes ten orders of monocots and two families of monocots (Petrosaviaceae and Dasypogonaceae) not yet assigned to any order. More recently, the Petrosaviaceae has been included in the Petrosaviales, and placed near the lilioid orders.[7] The family Hydatellaceae, assigned to order Poales in the APG II system, has since been recognized as being misplaced in the monocots, and instead proves to be most closely related to the water lilies, family Nymphaeaceae.

clade monocots :

- order Acorales
- order Alismatales
- order Asparagales
- order Dioscoreales
- order Liliales
- order Pandanales
- order Petrosaviales
- clade commelinids:
    - family Dasypogonaceae
- order Arecales
- order Commelinales
- order Poales
- order Zingiberales

The current phylogeny and composition of the monocots.[8]

Evolution

For a very long time, fossils of palm trees were believed to be the oldest monocots, first appearing 90 million years ago, but this estimate may not be entirely true (reviewed in Herendeen and Crane, 1995[9]). At least some putative monocot fossils have been found in strata as old as the eudicots (reviewed in Herendeen *et al.*, 1995[10]). The oldest fossils that are unequivocally monocots are pollen from the Late Barremian-Aptian - Early Cretaceous period, about 120-110 million years ago, and are assignable to clade-Pothoideae-Monstereae Araceae; being Araceae, sister to other Alismatales (Friis *et al.*, 2004:[11]) for fossil monocots, see Gandolfo et al. 2000 [71] and Friis et al. 2006b [72]). They have also found flower fossils of Triuridaceae (Pandanales) in Upper Cretaceous rocks in New Jersey (Gandolfo et al. 2002 [73]), becoming the oldest known sighting of saprophytic /mycotrophic habits in angiosperm plants and among the oldest known fossils of monocotyledons.

Grass sprouting on left (a monocot), showing hypogeal development (the cotyledon remains underground and is not visible). Compare to a dicot (right)

Topology of the angiosperm phylogenetic tree could infer that the monocots would be among the oldest lineages of angiosperms, which would support the theory that they are just as old as the eudicots. The pollen of the eudicots dates back 125 million years, so the lineage of monocots should be that old too.

The molecular age estimates also hold the view that monocots are as old as eudicots. Bremer (2000 [74] 2002 [75]), using the rbcL sequence data and the method of the middle way ("mean-path lengths method") to estimate divergence times, dates back the origin of the crown group of the monocots (the time *Acorus* genus divides the rest of the group) to some 134 million years, which would mean that estimates of the main group of monocots are even older. However Wikström et al. (2001 [76]), using Sanderson's (1997 [77]) non-parametric approach (nonparametric rate smoothing approach "), produced ages for the crown group of monocots between 158 and 141 million years up until now (see Sanderson et al. 2004 [78]), ages markedly older than Bremer's, so that the trunk group of the monocots would also be older than Bremer's estimations. The discrepancy between these two estimates is probably due to the highly conservative calibration point used in the study of Wikström et al. 2001 [76] (the split between Fagales and Cucurbitales were considered to be in the late Santonian period).

In fact the age of the monocot crown has been variously estimated, besides the two mentioned, it has been estimated around 200 ± 20 million years BC (Savard et al. 1994 [79]), 160 ± 16 million years BC (Goremykin et al. 1997 [80]), 135-131 million (Leebens-Mack et al. 2005 [81]), 133.8 to 124 million (Moore et al. 2007 [82]) , etc.

Assuming Triuridaceae is a member of Pandanales, their fossils would give support to a crown group age closer to Bremer's estimations (2000 [74]).

Bremer's estimation (2000 [74]) was used in a more recent analysis that formed the basis for dating the age of the monocots in general (Janssen and Bremer 2004 [83]).

The age of the core group of so-called 'nuclear monocot' or 'core monocots' by the APW ("core monocots" in English), which correspond to all orders except Acorales and Alismatales, is about 131 million years to present, and crown group age is about 126 million years to the present. The subsequent branching in this part of the tree (i.e., Petrosaviaceae, Dioscoreales + Pandanales and Liliales clades appeared), including the crown Petrosaviaceae group may be in the period around 125-120 million years BC (about 111 million years so far in Bremer 2000 [74]), and stem groups of all other orders, including Commelinadae would have diverged about or shortly after 115 million years (Janssen and Bremer 2004 [83]). These and many clades within these orders may have originated in southern Gondwana, i.e., Antarctica, Australasia, and southern South America (Bremer and Jansen 2006 [84]).

The aquatic monocot Alismatales have commonly been regarded as "primitive" (Hallier, 1905, [85] Arber 1925, [86] Hutchinson, 1934, [87] Cronquist 1968, [88] 1981, [5] Takhtajan 1969, [89] 1991 [90] Stebbins 1974, [91] Thorne 1976 [92]). They have also been considered to have the most primitive foliage, which were cross-linked as Dioscoreales (Dahlgren et al. 1985 [15] and Melanthiales (Thorne 1992a, [93] 1992b [94]). Keep in mind that, as stressed by Soltis et al. 2005, the "most primitive" monocot is not necessarily "the sister of everyone else." This is because the ancestral or primitive characters are inferred by means of the reconstruction of characteristic states, with the help of the phylogenetic tree. So primitive characters of monocots may be present in some derived groups. On the other hand, the basal taxa may exhibit many morphological autapomorphies. So although Acoraceae is the sister group to the remaining monocotyledons, the result does not imply that Acoraceae is "the most primitive monocot" in terms of its characteristics. In fact, Acoraceae is highly derived in most morphological characteristics, which is precisely why so many Alismatales Acoraceae occupied relatively imitative positions in trees produced by Chase et al. 1995b [14] and Stevenson and Loconte 1995. [50] (see section phylogeny).

Some authors support the idea of an aquatic phase as the origin of monocots (Henslow 1893, [95] and also cited and argued in the phylogeny section that Alismatales are the most primitive). The phylogenetic position of Alismatales (many water), which occupy a relationship with the rest except the Acoraceae, do not rule out the idea, because it could be 'the most primitive monocots' but not 'the most basal'. The Atactostele stem, the long and linear leaves, the absence of secondary growth (see the biomechanics of living in the water), roots in groups instead of a single root branching (related to the nature of the substrate), including sympodial use, are consistent with a water source. However, while monocots were sisters of the aquatic Ceratophyllales, or their origin is related to the adoption of some form of aquatic habit, it would not help much to the understanding of how it evolved to develop their distinctive anatomical features: the monocots seem so different from the rest of angiosperms and it's difficult to relate their morphology, anatomy and development and those of broad-leaved angiosperms (e.g. Zimmermann and Tomlinson 1972; [96] Tomlinson 1995 [22]).

In the past, taxa which had petiolate leaves with reticulate venation were considered "primitive" within the monocots, because of its superficial resemblance to the leaves of dicotyledons. Recent work suggests that these taxa are sparse in the phylogenetic tree of monocots, such as fleshy fruited taxa (excluding taxa with aril seeds dispersed by ants), the two features would be adapted to conditions that evolved together regardless (Dahlgren and Clifford 1982; [97] Patterson and Givnish 2002, [98] Givnish et al. 2005b, [16] 2006b [99]). Among the taxa involved were *Smilax, Trillium* (Liliales), *Dioscorea* (Dioscoreales), etc. A number of these plants are vines that tend to live in shaded habitats for at least part of their lives, and may also have a relationship with their shapeless stomata (see Cameron and Dickison 1998 [100] for references on this last characteristic). Reticulate venation seems to have appeared at least 26 times in monocots, in fleshy fruits 21 times (sometimes lost later), and the two characteristics, though different, showed strong signs of a tendency to be good or bad in tandem, a phenomenon Givnish et al. (2005b, [16] 2006b [99]) described as "concerted convergence" ("coordinated convergence").

Notes

[1] http://cmsdata.iucn.org/downloads/2008rl_stats_tables_all.xls

[2] Peter H. Raven, Ray Franklin Evert & Susan E. Eichhorn. (2005) *Biology of Plants*, 7th ed., page 459

[3] Reed, Barbara (2008). *Plant cryopreservation a practical guide* (http://www.springerlink.com/content/u1gt1354336rr343/). New York: Springer. pp. 241. ISBN 978-0-387-72276-4. .

[4] Mark W. Chase (2004). "Monocot relationships: an overview" (http://www.amjbot.org/cgi/content/full/91/10/1645). *American Journal of Botany* **91** (10): 1645–1655. doi:10.3732/ajb.91.10.1645. PMID 21652314. .

[5] Radosevich, Steven R.; Holt, Jodie S.; Ghersa, Claudio (1997). *Weed ecology: implications for management* (http://books.google.com/?id=uK9R7N-QaJMC&pg=RA1-PA149). New York: J. Wiley. ISBN 0-471-11606-8. .

[6] Donoghue, Michael J. (2005). "Key innovations, convergence, and success: macroevolutionary lessons from plant phylogeny" (http://www.phylodiversity.net/donoghue/publications/MJD_papers/2005/149_MJD_Paleo05.pdf). *Paleobiology* **31**: 77. doi:10.1666/0094-8373(2005)031[0077:KICASM]2.0.CO;2. .

[7] Cantino, Philip D.; James A. Doyle, Sean W. Graham, Walter S. Judd, Richard G. Olmstead, Douglas E. Soltis, Pamela S. Soltis, & Michael J. Donoghue (2007). "Towards a phylogenetic nomenclature of *Tracheophyta*" (http://www.phylodiversity.net/donoghue/publications/MJD_papers/2007/164_Cantino_Taxon07.pdf) (PDF). *Taxon* **56** (3): E1–E44. .

[8] "An update of the Angiosperm Phylogeny Group classification for the orders and families of flowering plants: APG III". *Botanical Journal of the Linnean Society* **161** (2): 105–121. 2009. doi:10.1111/j.1095-8339.2009.00996.x.

[9] Herendeen, P. S.; Crane, P. R. (1995). "The fossil history of the monocotyledons". In Rudall, P., Cribb, P. J., Cutler, D. F. & C. J. Humphries. *Monocotyledons: systematics and evolution.*. London: Royal BOtanic Gardens Kew. pp. 1–21.

[10] Herendeen, P. S.; P. R. Crane & A. Drinnan (1995). *Fagaceous flowers, fruits, and cupules from the Campanian (Late Cretaceous) of Central Georgia, USA.* International Journal of Plant Sciences. **156**. pp. 93–116. JSTOR 2474901.

[11] Friis, E. M.; Pedersen, K. R., and Crane, P. R. (2004). "Araceae from the early Cretaceous of Portugal: Evidence on the emergence of monocotyledons". *Proceedings of the National Academy of Sciences* **101** (47): 16565–16570. doi:10.1073/pnas.0407174101. PMID 15546982.

References

- Jerrold I. Davis, Dennis W. Stevenson, Gitte Petersen, Ole Seberg, Lisa M. Campbell, John V. Freudenstein, Douglas H. Goldman, Christopher R. Hardy, Fabian A. Michelangeli, Mark P. Simmons, Chelsea D. Specht, Francisco Vergara-Silva & Maria Gandolfo (2004). "A phylogeny of the monocots, as inferred from *rbc*L and *atp*A sequence variation, and a comparison of methods for calculating jackknife and bootstrap values" (http://epmb.berkeley.edu/vfs/PIs/Specht-CD/web/SystBot.pdf). *Systematic Botany* **29** (3): 467–510. doi:10.1600/0363644041744365.

- Chase M. W., D. E. Soltis, P. S. Soltis, P. J. Rudall, M. F. Fay, W. J. Hahn, S. Sullivan, J. Joseph, M. Molvray, P. J. Kores, T. J. Givnish, K. J. Sytsma & J. C. Pires (2000). Higher-level systematics of the monocotyledons: An assessment of current knowledge and a new classification. In: K. L. Wilson & D. A. Morrison, eds. *Monocots: Systematics and Evolution.*. CSIRO, Melbourne. 3–16. ISBN 0-643-06437-0

External links

- Tree of Life Web Project: Monocotyledons (http://tolweb.org/tree?group=Monocotyledons&contgroup=Euangiosperms)

- "Numbers of threatened species by major groups of organisms (1996–2004)" (http://web.archive.org/web/20060927212300/http://www.redlist.org/info/tables/table1). International Union for Conservation of Nature and Natural Resources. Archived from the original (http://www.redlist.org/info/tables/table1) on 2006-09-27. Retrieved 2006-12-15.

- Monocots Plant Life Forms (http://www.plantlifeforms.com/Classes128/MONOCOT_MONOCOTYLEDONEAE_503902_128.aspx)

Flowering plant

<table>
<tr><td colspan="2" align="center">Flowering plants</td></tr>
<tr><td colspan="2" align="center">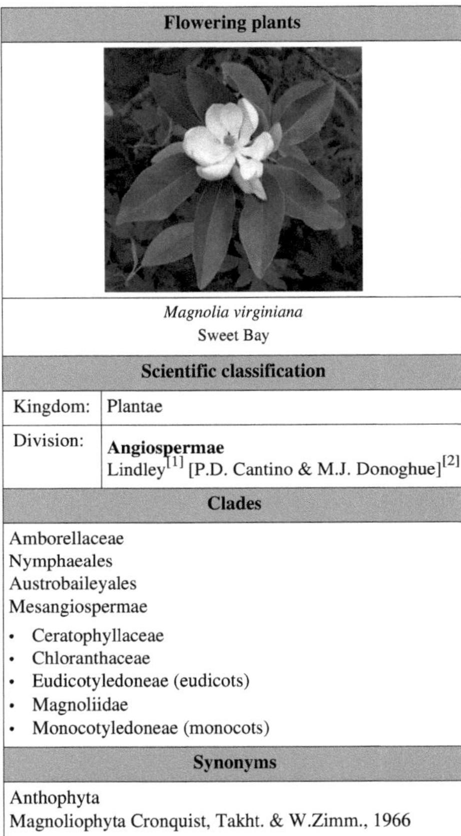</td></tr>
<tr><td colspan="2" align="center">Magnolia virginiana
Sweet Bay</td></tr>
<tr><td colspan="2" align="center">Scientific classification</td></tr>
<tr><td>Kingdom:</td><td>Plantae</td></tr>
<tr><td>Division:</td><td>Angiospermae
Lindley[1] [P.D. Cantino & M.J. Donoghue][2]</td></tr>
<tr><td colspan="2" align="center">Clades</td></tr>
<tr><td colspan="2">Amborellaceae
Nymphaeales
Austrobaileyales
Mesangiospermae

• Ceratophyllaceae
• Chloranthaceae
• Eudicotyledoneae (eudicots)
• Magnoliidae
• Monocotyledoneae (monocots)</td></tr>
<tr><td colspan="2" align="center">Synonyms</td></tr>
<tr><td colspan="2">Anthophyta
Magnoliophyta Cronquist, Takht. & W.Zimm., 1966</td></tr>
</table>

The **flowering plants (angiosperms)**, also known as **Angiospermae** or **Magnoliophyta**, are the most diverse group of land plants. Angiosperms are seed-producing plants like the gymnosperms and can be distinguished from the gymnosperms by a series of synapomorphies (derived characteristics). These characteristics include flowers, endosperm within the seeds, and the production of fruits that contain the seeds.

The ancestors of flowering plants diverged from gymnosperms around 245–202 million years ago, and the first flowering plants known to exist are from 140 million years ago. They diversified enormously during the Lower Cretaceous and became widespread around 100 million years ago, but replaced conifers as the dominant trees only around 60–100 million years ago.

Angiosperm derived characteristics

Bud of a pink rose

- Flowers

The flowers, which are the reproductive organs of flowering plants, are the most remarkable feature distinguishing them from other seed plants. Flowers aid angiosperms by enabling a wider range of adaptability and broadening the ecological niches open to them. This has allowed flowering plants to largely dominate terrestrial ecosystems.

- Stamens with two pairs of pollen sacs

Stamens are much lighter than the corresponding organs of gymnosperms and have contributed to the diversification of angiosperms through time with adaptations to specialized pollination syndromes, such as particular pollinators. Stamens have also become modified through time to prevent self-fertilization, which has permitted further diversification, allowing angiosperms eventually to fill more niches.

- Reduced male parts, three cells

The male gametophyte in angiosperms is significantly reduced in size compared to those of gymnosperm seed plants. The smaller pollen decreases the time from pollination — the pollen grain reaching the female plant — to fertilization. In gymnosperms, fertilization can occur up to a year after pollination, whereas in angiosperms, fertilization begins very soon after pollination. The shorter time leads to angiosperm plants' setting seeds sooner and faster than gymnosperms, which is a distinct evolutionary advantage.

- Closed carpel enclosing the ovules (carpel or carpels and accessory parts may become the fruit)

The closed carpel of angiosperms also allows adaptations to specialized pollination syndromes and controls. This helps to prevent self-fertilization, thereby maintaining increased diversity. Once the ovary is fertilized, the carpel and some surrounding tissues develop into a fruit. This fruit often serves as an attractant to seed-dispersing animals. The resulting cooperative relationship presents another advantage to angiosperms in the process of dispersal.

- Reduced female gametophyte, seven cells with eight nuclei

The reduced female gametophyte, like the reduced male gametophyte, may be an adaptation allowing for more rapid seed set, eventually leading to such flowering plant adaptations as annual herbaceous life-cycles, allowing the flowering plants to fill even more niches.

- Endosperm

In general, endosperm formation begins after fertilization and before the first division of the zygote. Endosperm is a highly nutritive tissue that can provide food for the developing embryo, the cotyledons, and sometimes the seedling when it first appears.

These distinguishing characteristics taken together have made the angiosperms the most diverse and numerous land plants and the most commercially important group to humans. The major exception to the dominance of terrestrial ecosystems by flowering plants is the coniferous forest.

Evolution

Further information: Evolutionary history of plants – Flowers

Fossilized spores suggest that higher plants (embryophytes) have lived on the land for at least 475 million years.[3] Early land plants reproduced sexually with flagellated, swimming sperm, like the green algae from which they evolved. An adaptation to terrestrialization was the development of upright meiosporangia for dispersal by spores to new habitats. This feature is lacking in the descendants of their nearest algal relatives, the Charophycean green algae. A later terrestrial adaptation took place with retention of the delicate, avascular sexual stage, the gametophyte, within the tissues of the vascular sporophyte. This occurred by spore germination within sporangia rather than spore release, as in non-seed plants. A current example of how this might have happened can be seen in the precocious spore germination in *Sellaginella*, the spike-moss. The result for the ancestors of angiosperms was enclosing them in a case, the seed. The first seed bearing plants, like the ginkgo, and conifers (such as pines and firs), did not produce flowers. The pollen grains (males) of *Ginkgo* and cycads produce a pair of flagellated, mobile sperm cells that "swim" down the developing pollen tube to the female and her eggs.

Flowers of *Malus sylvestris* (crab apple)

The apparently sudden appearance of relatively modern flowers in the fossil record initially posed such a problem for the theory of evolution that it was called an "*abominable mystery*" by Charles Darwin.[4] However, the fossil record has considerably grown since the time of Darwin, and recently discovered angiosperm fossils such as *Archaefructus*, along with further discoveries of fossil gymnosperms, suggest how angiosperm characteristics may have been acquired in a series of steps. Several groups of extinct gymnosperms, in particular seed ferns, have been proposed as the ancestors of flowering plants, but there is no continuous fossil evidence showing exactly how flowers evolved. Some older fossils, such as the upper Triassic *Sanmiguelia*, have been suggested. Based on current evidence, some propose that the ancestors of the angiosperms diverged from an unknown group of gymnosperms during the late Triassic (245–202 million years ago). A close relationship between angiosperms and gnetophytes, proposed on the basis of morphological evidence, has more recently been disputed on the basis of molecular evidence that suggest gnetophytes are instead more closely related to other gymnosperms.

The evolution of seed plants and later angiosperms appears to be the result of two distinct rounds of whole genome duplication events.[5] These occurred in 319 [6] million years ago and 192 [7] million years ago respectively.

The earliest known macrofossil confidently identified as an angiosperm, *Archaefructus liaoningensis*, is dated to about 125 million years BP (the Cretaceous period),[8] while pollen considered to be of angiosperm origin takes the fossil record back to about 130 million years BP. However, one study has suggested that the early-middle Jurassic plant *Schmeissneria*, traditionally considered a type of ginkgo, may be the earliest known angiosperm, or at least a close relative.[9] In addition, circumstantial chemical evidence has been found for the existence of angiosperms as early as 250 million years ago. Oleanane, a secondary metabolite produced by many flowering plants, has been found in Permian deposits of that age together with fossils of gigantopterids.[10] [11] Gigantopterids are a group of extinct seed plants that share many morphological traits with flowering plants, although they are not known to have been flowering plants themselves.

Recent DNA analysis based on molecular systematics [12] [13] showed that *Amborella trichopoda*, found on the Pacific island of New Caledonia, belongs to a sister group of the other flowering plants, and morphological studies [14] suggest that it has features that may have been characteristic of the earliest flowering plants.

The orders Amborellales, Nymphaeales, and Austrobaileyales diverged as separate lineages from the remaining angiosperm clade at a very early stage in flowering plant evolution.[15]

The great angiosperm radiation, when a great diversity of angiosperms appears in the fossil record, occurred in the mid-Cretaceous (approximately 100 million years ago). However, a study in 2007 estimated that the division of the five most recent (the genus *Ceratophyllum*, the family Chloranthaceae, the eudicots, the magnoliids, and the monocots) of the eight main groups occurred around 140 million years ago.[16] By the late Cretaceous, angiosperms appear to have dominated environments formerly occupied by ferns and cycadophytes, but large canopy-forming trees replaced conifers as the dominant trees only close to the end of the Cretaceous 65 millions years ago or even later, at the beginning of the Tertiary.[17] The radiation of herbaceous angiosperm occurred much later.[18] Yet, many fossil plants recognizable as belonging to modern families (including beech, oak, maple, and magnolia) appeared already at late Cretaceous.

It is generally assumed that the function of flowers, from the start, was to involve mobile animals in their reproduction processes. That is, pollen can be scattered even if the flower is not brightly colored or oddly shaped in a way that attracts animals; however, by expending the energy required to create such traits, angiosperms can enlist the aid of animals and, thus, reproduce more efficiently.

Two bees on a flower head of Creeping Thistle, *Cirsium arvense*

Island genetics provides one proposed explanation for the sudden, fully developed appearance of flowering plants. Island genetics is believed to be a common source of speciation in general, especially when it comes to radical adaptations that seem to have required inferior transitional forms. Flowering plants may have evolved in an isolated setting like an island or island chain, where the plants bearing them were able to develop a highly specialized relationship with some specific animal (a wasp, for example). Such a relationship, with a hypothetical wasp carrying pollen from one plant to another much the way fig wasps do today, could result in the development of a high degree of specialization in both the plant(s) and their partners. Note that the wasp example is not incidental; bees, which, it is postulated, evolved specifically due to mutualistic plant relationships, are descended from wasps.

Animals are also involved in the distribution of seeds. Fruit, which is formed by the enlargement of flower parts, is frequently a seed-dispersal tool that attracts animals to eat or otherwise disturb it, incidentally scattering the seeds it contains (see frugivory). While many such mutualistic relationships remain too fragile to survive competition and to spread widely, flowering proved to be an unusually effective means of reproduction, spreading (whatever its origin) to become the dominant form of land plant life.

Flower ontogeny uses a combination of genes normally responsible for forming new shoots.[19] The most primitive flowers are thought to have had a variable number of flower parts, often separate from (but in contact with) each other. The flowers would have tended to grow in a spiral pattern, to be bisexual (in plants, this means both male and female parts on the same flower), and to be dominated by the ovary (female part). As flowers grew more advanced, some variations developed parts fused together, with a much more specific number and design, and with either specific sexes per flower or plant, or at least "ovary-inferior".

Flower evolution continues to the present day; modern flowers have been so profoundly influenced by humans that some of them cannot be pollinated in nature. Many modern, domesticated flowers used to be simple weeds, which sprouted only when the ground was disturbed. Some of them tended to grow with human crops, perhaps already having symbiotic companion plant relationships with them, and the prettiest did not get plucked because of their beauty, developing a dependence upon and special adaptation to human affection.[20]

A few palaeontologists have also come up with a theory that flowering plants, or angiosperms, might have evolved because of dinosaurs; in other words, they believe that dinosaurs "created" flowers. One of the theory's biggest proponents is Robert T. Bakker. He theorizes that herbivorous dinosaurs, with their eating habits, forced plants to find new ways to develop new adaptations, in order to avoid predation by herbivores.

Classification

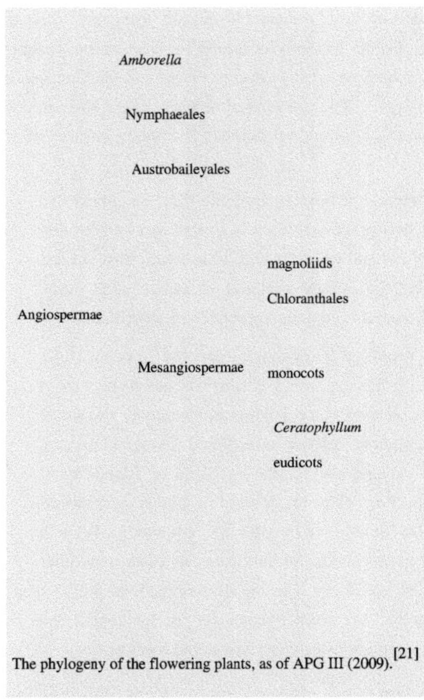

The phylogeny of the flowering plants, as of APG III (2009).[21]

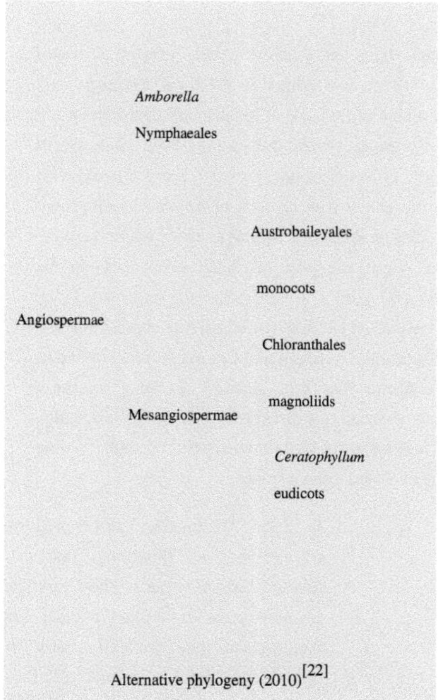

There are eight groups of living angiosperms:

- *Amborella* — a single species of shrub from New Caledonia
- Nymphaeales — about 80 species[23] — water lilies and Hydatellaceae
- Austrobaileyales — about 100 species[23] of woody plants from various parts of the world
- Chloranthales — several dozen species of aromatic plants with toothed leaves
- Magnoliidae — about 9,000 species,[23] characterized by trimerous flowers, pollen with one pore, and usually branching-veined leaves — for example magnolias, bay laurel, and black pepper
- Monocotyledonae — about 70,000 species,[23] characterized by trimerous flowers, a single cotyledon, pollen with one pore, and usually parallel-veined leaves — for example grasses, orchids, and palms
- *Ceratophyllum* — about 6 species[23] of aquatic plants, perhaps most familiar as aquarium plants
- Eudicotyledonae — about 175,000 species,[23] characterized by 4- or 5- merous flowers, pollen with three pores, and usually branching-veined leaves — for example sunflowers, petunia, buttercup, apples and oaks

The exact relationship between these eight groups is not yet clear, although there is agreement that the first three groups to diverge from the ancestral angiosperm were Amborellales, Nymphaeales, and Austrobaileyales.[24] The term basal angiosperms refers to these three groups. The five other groups form the clade Mesangiospermae. The relationship between the three largest of these groups (magnoliids, monocots and eudicots) remains unclear. Some analyses make the magnoliids the first to diverge, others the monocots.[22] *Ceratophyllum* seems to group with the eudicots rather than with the monocots.

History of classification

The botanical term "Angiosperm", from the Ancient Greek αγγείον, *angeíon* (receptacle, vessel) and σπέρμα, (seed), was coined in the form Angiospermae by Paul Hermann in 1690, as the name of that one of his primary divisions of the plant kingdom. This included flowering plants possessing seeds enclosed in capsules, distinguished from his Gymnospermae, or flowering plants with achenial or schizo-carpic fruits, the whole fruit or each of its pieces being here regarded as a seed and naked. The term and its antonym were maintained by Carolus Linnaeus with the same sense, but with restricted application, in the names of the orders of his class Didynamia. Its use with any approach to its modern scope became possible only after 1827, when Robert Brown established the existence of truly naked ovules in the Cycadeae and Coniferae, and applied to them the name Gymnosperms. From that time onward, as long as these Gymnosperms were, as was usual, reckoned as dicotyledonous flowering plants, the term Angiosperm was used antithetically by botanical writers, with varying scope, as a group-name for other dicotyledonous plants.

From 1736, an illustration of Linnaean classification

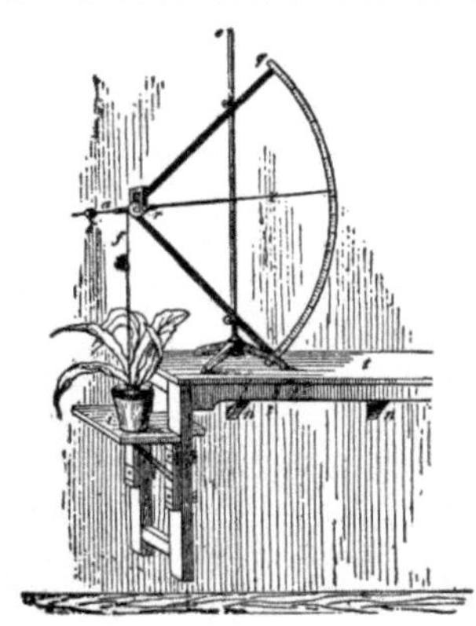

Auxanometer: Device for measuring increase or rate of growth in plants

In 1851, Hofmeister discovered the changes occurring in the embryo-sac of flowering plants, and determined the correct relationships of these to the Cryptogamia. This fixed the position of Gymnosperms as a class distinct from Dicotyledons, and the term Angiosperm then gradually came to be accepted as the suitable designation for the whole of the flowering plants other than Gymnosperms, including the classes of Dicotyledons and Monocotyledons. This is the sense in which the term is used today.

In most taxonomies, the flowering plants are treated as a coherent group. The most popular descriptive name has been Angiospermae (Angiosperms), with Anthophyta ("flowering plants") a second choice. These names are not linked to any rank. The Wettstein system and the Engler system use the name Angiospermae, at the assigned rank of subdivision. The Reveal system treated flowering plants as subdivision Magnoliophytina (Frohne & U. Jensen ex Reveal, Phytologia 79: 70 1996), but later split it to Magnoliopsida, Liliopsida, and Rosopsida. The Takhtajan system and Cronquist system treat this group at the rank of division, leading to the name Magnoliophyta (from the family name Magnoliaceae). The Dahlgren system and Thorne system (1992) treat this group at the rank of class, leading to the name Magnoliopsida. The APG system of 1998, and the later 2003[25] and 2009[21] revisions, treat the flowering plants as a clade called angiosperms without a formal botanical name. However, a formal classification was published alongside the 2009 revision in which the flowering plants form the Subclass Magnoliidae.[26]

The internal classification of this group has undergone considerable revision. The Cronquist system, proposed by Arthur Cronquist in 1968 and published in its full form in 1981, is still widely used but is no longer believed to accurately reflect phylogeny. A consensus about how the flowering plants should be arranged has recently begun to emerge through the work of the Angiosperm Phylogeny Group (APG), which published an influential reclassification of the angiosperms in 1998. Updates incorporating more recent research were published as APG II in 2003[25] and as APG III in 2009.[21] [27]

Traditionally, the flowering plants are divided into two groups, which in the Cronquist system are called *Magnoliopsida* (at the rank of class, formed from the family name *Magnoliacae*) and *Liliopsida* (at the rank of class, formed from the family name *Liliaceae*). Other descriptive names allowed by Article 16 of the ICBN include *Dicotyledones* or *Dicotyledoneae*, and *Monocotyledones* or *Monocotyledoneae*, which have a long history of use. In English a member of either group may be called a *dicotyledon* (plural *dicotyledons*) and *monocotyledon* (plural *monocotyledons*), or abbreviated, as *dicot* (plural *dicots*) and *monocot* (plural *monocots*). These names derive from the observation that the dicots most often have two *cotyledons*, or embryonic leaves, within each seed. The monocots usually have only one, but the rule is not absolute either way. From a diagnostic point of view, the number of cotyledons is neither a particularly handy nor a reliable character.

Monocot (left) and dicot seedlings

Recent studies, as by the APG, show that the monocots form a monophyletic group (clade) but that the dicots do not (they are paraphyletic). Nevertheless, the majority of dicot species do form a monophyletic group, called the *eudicots* or *tricolpates*. Of the remaining dicot species, most belong to a third major clade known as the Magnoliidae, containing about 9,000 species. The rest include a paraphyletic grouping of primitive species known collectively as the basal angiosperms, plus the families Ceratophyllaceae and Chloranthaceae.

Flowering plant diversity

The number of species of flowering plants is estimated to be in the range of 250,000 to 400,000.[28] [29] [30] The number of families in APG (1998) was 462. In APG II[25] (2003) it is not settled; at maximum it is 457, but within this number there are 55 optional segregates, so that the minimum number of families in this system is 402. In APG III (2009) there are 415 families.[21]

The diversity of flowering plants is not evenly distributed. Nearly all species belong to the eudicot (75%), monocot (23%) and magnoliid (2%) clades. The remaining 5 clades contain a little over 250 species in total, i.e., less than 0.1% of flowering plant diversity, divided among 9 families.

The most diverse families of flowering plants, in their APG circumscriptions, in order of number of species, are:

A poster of twelve different species of flowers of the *Asteraceae* family

1. Asteraceae or Compositae (daisy family): 23,600 species[31]
2. Orchidaceae (orchid family): 22,075 species[31]
3. Fabaceae or Leguminosae (pea family): 19,400[31]
4. Rubiaceae (madder family): 13,150[32]
5. Poaceae or Gramineae (grass family): 10,035[31]
6. Lamiaceae or Labiatae (mint family): 7,173[31]
7. Euphorbiaceae (spurge family): 5,735[31]
8. Melastomataceae (melastome family): 5,005[31]
9. Myrtaceae (myrtle family): 4,620[31]
10. Apocynaceae (dogbane family): 4,555[31]

In the list above (showing only the 10 largest families), the Orchidaceae and Poaceae are monocot families; the others are eudicot families.

Vascular anatomy

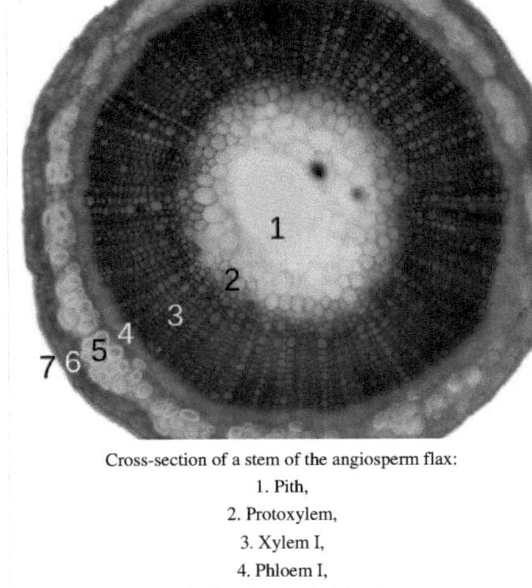

Cross-section of a stem of the angiosperm flax:
1. Pith,
2. Protoxylem,
3. Xylem I,
4. Phloem I,
5. Sclerenchyma (bast fibre),
6. Cortex,
7. Epidermis

The amount and complexity of tissue-formation in flowering plants exceeds that of gymnosperms. The vascular bundles of the stem are arranged such that the xylem and phloem form concentric rings.

In the dicotyledons, the bundles in the very young stem are arranged in an open ring, separating a central pith from an outer cortex. In each bundle, separating the xylem and phloem, is a layer of meristem or active formative tissue known as cambium. By the formation of a layer of cambium between the bundles (interfascicular cambium), a complete ring is formed, and a regular periodical increase in thickness results from the development of xylem on the inside and phloem on the outside. The soft phloem becomes crushed, but the hard wood persists and forms the bulk of the stem and branches of the woody perennial. Owing to differences in the character of the elements produced at the beginning and end of the season, the wood is marked out in transverse section into concentric rings, one for each season of growth, called annual rings.

Among the monocotyledons, the bundles are more numerous in the young stem and are scattered through the ground tissue. They contain no cambium and once formed the stem increases in diameter only in exceptional cases.

The flower, fruit, and seed

Flowers

The characteristic feature of angiosperms is the flower. Flowers show remarkable variation in form and elaboration, and provide the most trustworthy external characteristics for establishing relationships among angiosperm species. The function of the flower is to ensure fertilization of the ovule and development of fruit containing seeds. The floral apparatus may arise terminally on a shoot or from the axil of a leaf (where the petiole attaches to the stem). Occasionally, as in violets, a flower arises singly in the axil of an ordinary foliage-leaf. More typically, the flower-bearing portion of the plant is sharply distinguished from the foliage-bearing or vegetative portion, and forms a more or less elaborate branch-system called an inflorescence.

There are two kinds of reproductive cells produced by flowers. Microspores, which will divide to become pollen grains, are the "male" cells and are borne in the stamens (or microsporophylls). The "female" cells called megaspores, which will divide to become the egg-cell (megagametogenesis), are contained in the ovule and enclosed in the carpel (or megasporophyll).

The flower may consist only of these parts, as in willow, where each flower comprises only a few stamens or two carpels. Usually, other structures are present and serve to protect the sporophylls and to form an envelope attractive to pollinators. The individual members of these surrounding structures are known as sepals and petals (or tepals in flowers such as *Magnolia* where sepals and petals are not distinguishable from each

A collection of flowers forming an inflorescence

other). The outer series (calyx of sepals) is usually green and leaf-like, and functions to protect the rest of the flower, especially the bud. The inner series (corolla of petals) is, in general, white or brightly colored, and is more delicate in structure. It functions to attract insect or bird pollinators. Attraction is effected by color, scent, and nectar, which may be secreted in some part of the flower. The characteristics that attract pollinators account for the popularity of flowers and flowering plants among humans.

While the majority of flowers are perfect or hermaphrodite (having both male and female parts in the same flower structure), flowering plants have developed numerous morphological and physiological mechanisms to reduce or prevent self-fertilization. Heteromorphic flowers have short carpels and long stamens, or vice versa, so animal pollinators cannot easily transfer pollen to the pistil (receptive part of the carpel). Homomorphic flowers may employ a biochemical (physiological) mechanism called self-incompatibility to discriminate between self- and non-self pollen grains. In other species, the male and female parts are morphologically separated, developing on different flowers.

Fertilization and embryogenesis

Double fertilization refers to a process in which two sperm cells fertilize cells in the ovary. This process begins when a pollen grain adheres to the stigma of the pistil (female reproductive structure), germinates, and grows a long pollen tube. While this pollen tube is growing, a haploid generative cell travels down the tube behind the tube nucleus. The generative cell divides by mitosis to produce two haploid (n) sperm cells. As the pollen tube grows, it makes its way from the stigma, down the style and into the ovary. Here the pollen tube reaches the micropyle of the ovule and digests its way into one of the synergids, releasing its contents (which include the sperm cells). The synergid that the cells were released into degenerates and one sperm makes its way to fertilize the egg cell, producing a diploid ($2n$) zygote. The second sperm cell fuses with both central cell nuclei, producing a triploid ($3n$) cell. As the zygote develops into an embryo, the triploid cell develops into the endosperm, which serves as the embryo's food supply. The ovary now will develop into fruit and the ovule will develop into seed.

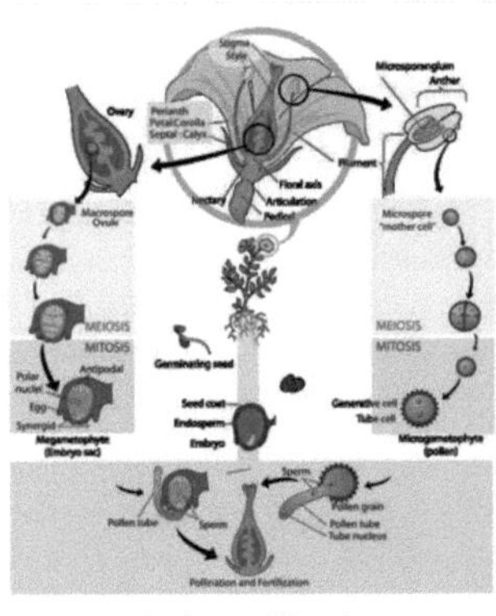

Angiosperm life cycle

Fruit and seed

The fruit of the *Aesculus* or Horse Chestnut tree

As the development of embryo and endosperm proceeds within the embryo-sac, the sac wall enlarges and combines with the nucellus (which is likewise enlarging) and the integument to form the *seed-coat*. The ovary wall develops to form the fruit or pericarp, whose form is closely associated with the manner of distribution of the seed.

Frequently, the influence of fertilization is felt beyond the ovary, and other parts of the flower take part in the formation of the fruit, e.g., the floral receptacle in the apple, strawberry, and others.

The character of the seed-coat bears a definite relation to that of the fruit. They protect the embryo and aid in dissemination; they may also directly promote germination. Among plants with indehiscent fruits, in general, the fruit provides protection for the embryo and secures dissemination. In this case, the seed-coat is only slightly developed. If the fruit is dehiscent and the seed is exposed, in general, the seed-coat is well developed, and must discharge the functions otherwise executed by the fruit.

Economic importance

Agriculture is almost entirely dependent upon angiosperms, which also provide a significant amount of livestock feed. Of all the families plants, the Poaceae, or grass family, is by far the most important, providing the bulk of all feedstocks (rice, corn — maize, wheat, barley, rye, oats, pearl millet, sugar cane, sorghum). The Fabaceae, or legume family, comes in second place. Also of high importance are the Solanaceae, or nightshade family (potatoes, tomatoes, and peppers, among others), the Cucurbitaceae, or gourd family (also including pumpkins and melons), the Brassicaceae, or mustard plant family (including rapeseed and the innumerable varieties of the cabbage species *Brassica oleracea*), and the Apiaceae, or parsley family. Many of our fruits come from the Rutaceae, or rue family (including oranges, lemons, grapefruits, etc.), and the Rosaceae, or rose family (including apples, pears, cherries, apricots, plums, etc.).

In some parts of the world, certain single species assume paramount importance because of their variety of uses, for example the coconut (*Cocos nucifera*) on Pacific atolls, and the olive (*Olea europaea*) in the Mediterranean region.

Flowering plants also provide economic resources in the form of wood, paper, fiber (cotton, flax, and hemp, among others), medicines (digitalis, camphor), decorative and landscaping plants, and many other uses. The main area in which they are surpassed by other plants is timber production.

See also

- List of garden plants
- List of plants by common name
- List of plant orders
- List of systems of plant taxonomy

References

[1] Lindley, J (1830). *Introduction to the Natural System of Botany.* London: Longman, Rees, Orme, Brown, and Green. xxxvi.

[2] Cantino, Philip D.; James A. Doyle, Sean W. Graham, Walter S. Judd, Richard G. Olmstead, Douglas E. Soltis, Pamela S. Soltis, & Michael J. Donoghue (2007). "Towards a phylogenetic nomenclature of *Tracheophyta*". *Taxon* **56** (3): E1–E44.

[3] Edwards, D (2000). "The role of mid-palaeozoic mesofossils in the detection of early bryophytes". *Philos Trans R Soc Lond B Biol Sci* **355** (1398): 733–755. doi:10.1098/rstb.2000.0613. PMC 1692787. PMID 10905607.

[4] Davies, T. J. (2004). "Darwin's abominable mystery: Insights from a supertree of the angiosperms". *Proceedings of the National Academy of Sciences* **101** (7): 1904–9. doi:10.1073/pnas.0308127100. PMC 357025. PMID 14766971.

[5] Jiao, Yuannian; Wickett, Norman J.; Ayyampalayam, Saravanaraj; Chanderbali, André S.; Landherr, Lena; Ralph, Paula E.; Tomsho, Lynn P.; Hu, Yi et al (2011). "Ancestral polyploidy in seed plants and angiosperms". *Nature* **473** (7345): 97–100. doi:10.1038/nature09916. PMID 21478875.

[6] http://toolserver.org/~verisimilus/Timeline/Timeline.php?Ma=319

[7] http://toolserver.org/~verisimilus/Timeline/Timeline.php?Ma=192

[8] Sun G., Ji Q., Dilcher D.L., Zheng S., Nixon K.C., Wang X. (2002). "Archaefructaceae, a New Basal Angiosperm Family" (http://www.sciencemag.org/cgi/content/abstract/296/5569/899?ck=nck&siteid=sci&ijkey=8dZ6zTqF606ps&keytype=ref). *Science* **296** (5569): 899–904. doi:10.1126/science.1069439. PMID 11988572. .

[9] Xin Wing; Shuying Duan, Baoyin Geng, Jinzhong Cui and Yong Yang (2007). "Schmeissneria: A missing link to angiosperms?". *BMC Evolutionary Biology* **7**: 14. doi:10.1186/1471-2148-7-14. PMC 1805421. PMID 17284326.

[10] Taylor, David Winship; Li, Hongqi; Dahl, Jeremy; Fago, Fred J.; Zinniker, David; Moldowan, J. Michael (2006). "Biogeochemical evidence for the presence of the angiosperm molecular fossil oleanane in Paleozoic and Mesozoic non-angiospermous fossils". *Paleobiology* **32** (2): 179. doi:10.1666/0094-8373(2006)32[179:BEFTPO]2.0.CO;2. ISSN 0094-8373.

[11] Oily Fossils Provide Clues To The Evolution Of Flowers (http://www.sciencedaily.com/releases/2001/04/010403071438.htm) — ScienceDaily (Apr. 5, 2001)

[12] NOVA — Transcripts — First Flower (http://www.pbs.org/wgbh/nova/transcripts/3405_flower.html) — PBS Airdate: April 17, 2007

[13] Soltis, D. E.; Soltis, P. S. (2004). "*Amborella* not a "basal angiosperm"? Not so fast". *American Journal of Botany* **91** (6): 997–1001. doi:10.3732/ajb.91.6.997. PMID 21653455.

[14] South Pacific plant may be missing link in evolution of flowering plants (http://www.eurekalert.org/pub_releases/2006-05/uoca-spp051506.php) — Public release date: 17-May-2006

[15] Vialette-Guiraud, AC; Alaux, M; Legeai, F; Finet, C; Chambrier, P; Brown, SC; Chauvet, A; Magdalena, C et al (2011). "Cabomba as a model for studies of early angiosperm evolution". *Annals of botany* **108** (4): 589–98. doi:10.1093/aob/mcr088. PMC 3170152. PMID 21486926.

[16] Moore, M. J.; Bell, C. D.; Soltis, P. S.; Soltis, D. E. (2007). "Using plastid genome-scale data to resolve enigmatic relationships among basal angiosperms". *Proceedings of the National Academy of Sciences* **104** (49): 19363–8. doi:10.1073/pnas.0708072104. PMC 2148295. PMID 18048334.

[17] David Sadava; H. Craig Heller; Gordon H. Orians; William K. Purves, David M. Hillis (December 2006). *Life: the science of biology* (http://books.google.com/books?id=1m0_FLEjd-cC&pg=PA477). Macmillan. pp. 477–. ISBN 9780716776741. . Retrieved 4 August 2010.

[18] Stewart, Wilson Nichols; Rothwell, Gar W. (1993). *Paleobotany and the evolution of plants* (2nd ed.). Cambridge Univ. Press. p. 498. ISBN 0521233151.

[19] Age-Old Question On Evolution Of Flowers Answered (http://unisci.com/stories/20012/0615015.htm) — 15-Jun-2001

[20] Human Affection Altered Evolution of Flowers (http://www.livescience.com/othernews/050526_flower_power.html) — By Robert Roy Britt, LiveScience Senior Writer (posted: 26 May 2005 06:53 am ET)

[21] Angiosperm Phylogeny Group (2009). "An update of the Angiosperm Phylogeny Group classification for the orders and families of
       flowering plants: APG III" (http://www3.interscience.wiley.com/journal/122630309/abstract). *Botanical Journal of the Linnean Society*
       161 (2): 105–121. doi:10.1111/j.1095-8339.2009.00996.x. . Retrieved 2010–12–10.
[22] Bell, C.D.; Soltis, D.E. & Soltis, P.S. (2010). "The Age and Diversification of the Angiosperms Revisited". *American Journal of Botany* **97**
       (8): 1296–1303. doi:10.3732/ajb.0900346. PMID 21616882., p. 1300
[23] Jeffrey D. Palmer, Douglas E. Soltis and Mark W. Chase, Chase, M. W. (2004). *Figure 2* (http://www.amjbot.org/cgi/content/full/91/
       10/1437/F2). "The plant tree of life: an overview and some points of view" (http://www.amjbot.org/cgi/content/full/91/10/1437).
       American Journal of Botany **91** (10): 1437–1445. doi:10.3732/ajb.91.10.1437. PMID 21652302. .
[24] Pamela S. Soltis and Douglas E. Soltis (2004). "The origin and diversification of angiosperms" (http://www.amjbot.org/cgi/content/full/
       91/10/1614). *American Journal of Botany* **91** (10): 1614–1626. doi:10.3732/ajb.91.10.1614. PMID 21652312. .
[25] Angiosperm Phylogeny Group (2003). "An update of the Angiosperm Phylogeny Group classification for the orders and families of
       flowering plants: APG II" (http://www.blackwell-synergy.com/links/doi/10.1046/j.1095-8339.2003.t01-1-00158.x/full/). *Botanical
       Journal of the Linnean Society* **141** (4): 399–436. doi:10.1046/j.1095-8339.2003.t01-1-00158.x. .
[26] Chase, Mark W. & Reveal, James L. (2009). "A phylogenetic classification of the land plants to accompany APG III". *Botanical Journal of
       the Linnean Society* **161** (2): 122–127. doi:10.1111/j.1095-8339.2009.01002.x. | Chase & Reveal 2009
[27] *As easy as APG III - Scientists revise the system of classifying flowering plants* (http://www.linnean.org/index.php?id=448). The Linnean
       Society of London. 2009-10-08. . Retrieved 2009-10-29.
[28] Thorne, R. F. (2002). "How many species of seed plants are there?" (http://www.ingentaconnect.com/content//iapt/tax/2002/
       00000051/00000003/art00009). *Taxon* **51** (3): 511–522. doi:10.2307/1554864. JSTOR 1554864. .>
[29] Scotland, R. W. & Wortley, A. H. (2003). "How many species of seed plants are there?" (http://www.ingentaconnect.com/content/iapt/
       tax/2003/00000052/00000001/art00011). *Taxon* **52** (1): 101–104. doi:10.2307/3647306. JSTOR 3647306. .
[30] Govaerts, R.url=http://www.ingentaconnect.com/content/iapt/tax/2003/00000052/00000003/art00016 (2003). "How
       many species of seed plants are there? — a response". *Taxon* **52** (3): 583–584. doi:10.2307/3647457. JSTOR 3647457.
[31] Stevens, P.F. (2001 onwards). "Angiosperm Phylogeny Website (at Missouri Botanical Garden)" (http://www.mobot.org/MOBOT/
       Research/APweb/welcome.html). .
[32] "Kew Scientist 30 (October2006)" (http://www.kew.org/kewscientist/ks_30.pdf). .

Further reading

- Cronquist, Arthur (1981). *An Integrated System of Classification of Flowering Plants*. New York: Columbia
  Univ. Press. ISBN 0-23-103880-1.
- Heywood, V. H., Brummitt, R. K., Culham, A. & Seberg, O. (2007). *Flowering Plant Families of the World*.
  Richmond Hill, Ontario, Canada: Firefly Books. ISBN 1-55407-206-9.
- Dilcher, D. (2000). "Toward a new synthesis: Major evolutionary trends in the angiosperm fossil record".
  Proceedings of the National Academy of Sciences **97** (13): 7030. doi:10.1073/pnas.97.13.7030.
- Simpson, M.G. *Plant Systematics*, 2nd Edition. Elsevier/Academic Press. 2010.
- Raven, P.H., R.F. Evert, S.E. Eichhorn. *Biology of Plants*, 7th Edition. W.H. Freeman. 2004.

External links

- Cole, Theodor C.H.; Hilger, Dr. Harmut H. Angiosperm Phylogeny Poster – Flowering Plant Systamatics (http://
  www.biologie.fu-berlin.de/sysbot/poster/poster1.pdf)
- Cromie, William J. (December 16, 1999). "Oldest Known Flowering Plants Identified By Genes" (http://www.
  news.harvard.edu/gazette/1999/12.16/angiosperms.html). Harvard University Gazette.
- Watson, L. and Dallwitz, M.J. (1992 onwards). The families of flowering plants: descriptions, illustrations,
  identification, information retrieval. (http://www.biologie.uni-hamburg.de/b-online/delta/angio/)

This article incorporates text from a publication now in the public domain: Chisholm, Hugh, ed. (1911).
Encyclopædia Britannica (11th ed.). Cambridge University Press.

Article Sources and Contributors

Semiarundinaria *Source*: http://en.wikipedia.org/w/index.php?title=Semiarundinaria *Contributors*: Anna Frodesiak, D6, Flakinho, Hesperian, IKAllmer, MPF, Mmcknight4, Qwertzy2, Visik

Shibataeinae *Source*: http://en.wikipedia.org/w/index.php?title=Shibataeinae *Contributors*: Cdogsimmons, EncycloPetey, GTBacchus, Hesperian, Manticore, Mikegodwin, Mmcknight4, Qwertzy2, Waacstats, 1 anonymous edits

Bamboo *Source*: http://en.wikipedia.org/w/index.php?title=Bamboo *Contributors*: 24fan24, 2D, 2help, 2over0, 49erInOregon, 78.26, 7thell, 919tswells, AManWithNoPlan, AMuseo, Abductive, Acalamari, Acroterion, Adolphus79, Afterrock81, Ahoerstemeier, Aitias, Akiha Tohno, Alan Liefting, Alansohn, Aleenf1, Alefeb, AlexanderKaras, Alexius08, Alexthusy, Allstarecho, Alton, Ambientbp, Anaxial, Andre Engels, AndrewDressel, Andryono, Andycjp, Angelofdeath275, Anna Frodesiak, Antandrus, Argo Navis, Arima, Arong53, Arpingstone, Art LaPella, AshLin, Assm, Avono, AxelBoldt, Az1568, Azcolvin429, Azza-bazoo, Baa, Badagnani, Bahudhara, Bamboo Massage, BambooSelect, Bambouhabitat, Bartledan, Bathrobe, Battoe19, Bawtyshouse, BazookaJoe, Bcore, Bdentremont, Beeblebrox, Beethoven's Last Nightmare, Beland, Belovedfreak, Ben Ben, Bensmith53, Bgag, BiT, Big55e, BigMoose22, Bigyabbie, BillWSmithJr, Billinghurst, Bit, Blazingbamboo, BlckKnght, Bnyup, Bob Burkhardt, Bobo192, Bongwarrior, Borok, Brianga, Bruno Cezar Bastos Ruano, Brunton, Brutaldeluxe, Brutannica, Bubbha, Buck09, Buckyboy314, Burntsauce, Burschik, CCNichole2015, CL, Cacahuate, Caeruleancentaur, Calmer Waters, Cameron mcroberts, Can't sleep, clown will eat me, CanadianLinuxUser, CanadianNine, Capricorn42, Castlan, Ccrrccrr, Chaipau, Chanjongjung, Chicbicyclist, Chowbok, Chris 73, Chris j wood, ChromeWire, Chuljin, Cjsannicolas, Claire French, Clariosophic, Cliff smith, Cmtigger, Cngodles, Coelacan, Cold Season, CommonsDelinker, Conscious, Cosmotron, Crazytales, Cst17, Curtholr, Cyde, DARTH SIDIOUS 2, DHN, DMacks, DVD R W, DaGizza, Dalampasigan, DandyAndyy, DanielCD, Danielt, Danio, Dar25ren, Dark Dragon Sword, DarkRenji, DaughterofSun, David Merrill, Dcattell, DeadEyeArrow, DeadLennon, Deadkid dk, Deagle AP, Delicious carbuncle, Delnatura, Deor, DerHexer, Dfgriggs, Dhaluza, Didup, Dillard421, Dirie, Dirkbb, Discospinster, Djlayton4, DocWatson42, Doniago, DougsTech, DrJGMD, Dragons953, Drake144, DreamGuy, Drilnoth, Drkmsoni, Drunauthorized, Ds13, Dunkyj, Dureo, Durova, Dust Filter, Dyepoy05, Dyl, E2eamon, Ebarcell, Edison, Edward321, Eequor, Ehn, EllePollack, EmjayMiller, Environmentalist and Historian, Equalshot, Eric-Wester, Erwinmeyer, Euchiasmus, Evan-Amos, Eyu100, Falcon8765, Faseidman, Fastily, Favonian, Felcorm, Femto, Fenderman416, Fir0002, Firsfron, Flashjt85, Fourthords, Fran McCrory, Fredrik, FreitagNazi, FreplySpang, Func, GBI 321, GMcGath, GTBacchus, Gadren, Gaius Cornelius, GamblinMonkey, Gentgeen, George Church, George Leung, Geosak, Gilliam, Gimmetrow, Gimpel, Gjl, GlassAcorn, Glenn, Goblue8888, Godfrey Daniel, Gogo Dodo, GoingBatty, Graham87, GreatWhiteNortherner, GregorB, Grenavitar, Grinchitude, Guadua, Guest9999, Gunkarta, Hadal, Haham hanuka, Hanacy, Hankwang, Hanwen, HenryLi, Henryiscool, Heron, Herr Beethoven, Hersfold, Hesperian, Hibernian, Hikaru.dream, Hillarychen, Hongthay, Horrorshowj, Hu Gadarn, Hunnjazal, II MusLiM HyBRiD II, IRP, Iamwisesun, Iancarter, Icez, Ichwan Palongengi, Idont Havaname, Imc, Indeterminate, Irishguy, Isogolem, J.delanoy, JForget, JNW, JaGa, Jabam, Jackhynes, Jaegerzewish, Jak123, Jake Wartenberg, Jan eissfeldt, Janrpeters, Jaranda, Jay Bogle, Jebus989, Jeff G., Jeffmcneill, JeramieHicks, Jerky duude, Jidanni, Jimmy Hammerfist, Jj137, Jjscientist, Jmundo, JoJan, Joaobambu, John Hill, JohnSankey, Johnbod, Joinred1127, Jojalozzo, Jojit fb, Jont76, Joofroman, JorisvS, Jose77, Josh Parris, Jpark3909, Jrh7925, Jsbisa, Juhachi, Julia Rossi, Julian Morrison, Juliancolton, Jusdafax, Jwoodger, K3kanyi, KFilz, Ka Faraq Gatri, Kandar, Kapilthakur79, KarlM, Katiegoats, Keiyulin, Kerrow, Khatores, Khoikhoi, Kilopi, KindGoat, Kingdon, Kintetsubuffalo, Kjrajesh, Kleinhev, Klewung, Koolbamboo, Krashlandon, Kricket, Ksyrie, Kubigula, Kudzu9, Kutu62, Kw0134, Kwamikagami, Kyaw 2003, L Kensington, LDHan, LaMenta3, Lady Tenar, Langbein Rise, Langra, Laurent paris, Lawrencekhoo, Lear's Fool, Legotech, Lemurbaby, Leolaursen, Leuko, Leyo, Liam, Liaocyed, LilHelpa, LinguistAtLarge, Logan, Look2See1, Loren36, Luv yuol, M.wyatt.cole, MFdeS, MONGO, MPF, Macboots, MacedonianBoy, Madger, Madman, Madsy12, Magnus Manske, Mambg, Mandarax, Mani1, Mapleowns23, Marek69, Mark Richards, Markussss1, Marshman, Master Scott Hall, Master of the Oríchalcos, Matahari Pagi, Mateo SA, Materialscientist, Mathiasjesus, MaxEspinho, Mc9205, McSly, Mctrain, Madd4696, Mediadar, Mentifisto, Merbabu, Mhockey, Mickg123, MidareIwashi, Mieciu K, Mikeblas, Mikeo, Mikm, Milnivlek, Missionviper77, MisterSheik, Mmyotis, Modify, Moeng, Mohylek, Molarty, Mongol, MonoAV, Mp3digitalplayer123, Mpm2121, MrDarwin, MrFloatingIP, Mrflowerpot, Mrosen814, Mtking, MuVo100, Murtasa, N96, Nakon, Namazu-tron, Nameless123456, Naniwako, Naohiro19 revertvandal, Neep, Nehrams2020, Neko-chan, Neoncow, NerdyScienceDude, Neutrality, NewEnglandYankee, Neweco, Ngorongoro, Niduzzi, Nihonjoe, Nikki311, Nima Baghaei, Noctibus, Nomadhacker, Nono64, NotALizard, Noveltyghost, Nposs, Ntsimp, NuclearWarfare, Oda Mari, Ohnoitsjamie, OlEnglish, Olivier, Olivierd, OllieFury, OneWorldMuso, Oobopshark, Opus88888, Orinhardy, Ottawa4ever, Ovis23, Oxymoron83, PBS-AWB, PDH, PRiis, Pakkanar, Param2, Pauly1060, Pawyilee, Pax:Vobiscum, Payo1, Pekinensis, PericlesofAthens, Peruvianllama, Pevarnj, Pgan002, Philip Trueman, PhilipDM, Photon87, Phr, Piano non troppo, PierreAbbat, Pinethicket, Pkmnnrd4evr, Pluma, Pmarshal, Pnkrockr, Portillo, PotatoSamurai, Powertothepeople12345, Praseprase, Priyatpr, Prof. Ofallofit, Prumpf, Prunesqualer, Pstiennon, PubliusVP, Pufferfish101, Puntori, Quebec99, Quintote, Quirky, Qwertzy2, Qwfp, R'n'B, R45, RMB1987, RW Marloe, Radiant chains, Rajesh dangi, RandomAct, Ratiocinate, Rattandy, Ravidreams, Raywil, Razirazo 90, Recognizance, Rehmano123, Rekrutacja, Remilo, Renato Caniatti, Reptiles, Revth, RexNL, ReyBrujo, Ricardo4max, RicegoneWILD, Rich Farmbrough, Richard Chesler, Richkoi, Ridyu, Rjanag, Rjd0060, Rjwilmsi, Rkitko, Robin S, Roche-Kerr, Roleplayer, Rominaguitar, Ronhjones, Rossami, Rrbe, Rror, Rsciaccio, RuiRobert, Runningonbrains, Rwflammang, SR, Saga City, Saintrain, Sam, SamErin117, Samikbhattacharya123, Samuel Grant, Sandeman684, Sander123, Sango123, Sapphirine, Saturn star, Sceptre, SchfiftyThree, Schwabea, Schwern, Schzmo, Scotteaux, Seaphoto, Seb az86556, Sedinalpac, SemblaceII, Semper discens, Sevilledade, Shalom Yechiel, Shelia258, Shirulashem, Shoeofdeath, Shoh Ueno., Shuffdog, Shyam, SiobhanHansa, Sjschen, Skepticus, Skizzik, SkyMachine, Smartse, SmellySmells, SmileToday, Smokizzy, Snigbrook, Snowmanradio, Socialservice, Somamary, Some jerk on the Internet, Spettro9, Spincr, Splitbamboo, SpuriousQ, Starvelez, SteinbDJ, Stephen G. Brown, StephenJ.Arnold, Steve Dufour, Steven Walling, Steven Weston, Stnz, StradivariusTV, Subcontinental, SunCreator, Susannelucas, Svick, Swanbau, Swaq, Swarm, T.santi, THEN WHO WAS PHONE?, TJ Fisher, TUF-KAT, TaintedMustard, Taonix (renamed), Tarquin, Tassedethe, Taylornator100, Tcp-ip, Tekwinigal, Teratornis, Teslatesla, Thamizhpparithi Maari, Tharish, The Thing That Should Not Be, TheLeopard, TheNothingNihilates, Thecheesykid, Theoden Eowyn, Thepixelvixen, Thingg, Thomas Larsen, Thparkth, Tide rolls, Tijuana Brass, Timtots, Tinchu, Tintenfischlein, Tmguchi, Tmoney15, Tmwerty, Tom Hulse, Tommizzzle, Tommy2010, Tonytnnt, Transity, Tropicalbamboo, Turmeric, Twalls, Tyberiuswp, TylerJacobson, Uncle Dick, Unschool, Urkelman, Useight, User A1, UserDoe, UtherSRG, VASANTH S.N., Vamooom, Van helsing, Vancouverbc, Vanhoabui, Vencel, Vernalautumn, Vina, Vinsfan368, Voodooom, Vozesdoalem, W Tanoto, WadeSimMiser, Ward3001, Wavelength, Wayne Slam, Weedgarden, Whitesandblacker, Wikid77, Wikievil666, Wikiramp, William Avery, Wimt, Wingman358, Wolfkeeper, Wolfmankurd, Woohookitty, WriterHound, Wtmitchell, Wwhat, X96lee15, Xandi94, Xezbeth, Xiao Li, Ximplo, Xp54321, Xufanc, Xxnightbarkxx, YPg4Gm, Yanksbuff, Ybarkai, Yoglok, Yuma en, Z.E.R.O., Zephyr2k, Zvn, Zzuuzz, יולי אורב, ‏.24 حمد‏.غ‏.دم, अमित भटनागर, 1413 anonymous edits

Bambusoideae *Source*: http://en.wikipedia.org/w/index.php?title=Bambusoideae *Contributors*: Alan Liefting, Andycjp, DanielCD, Deor, Drbreznjev, Flakinho, Gaius Cornelius, Hesperian, IKAllmer, Joanjoc, Look2See1, MPF, MiltonT, Mmcknight4, Mpm2121, Nipisiquit, Penarc, PuzzletChung, Qwertzy2, Remilo, Scott182, SimonP, UtherSRG, Vuong Ngan Ha, Woohookitty, 12 anonymous edits

Poaceae *Source*: http://en.wikipedia.org/w/index.php?title=Poaceae *Contributors*: 12345benz, 1984, Aelwyn, Alan Liefting, Alexei Kouprianov, Alvis, Anthony Appleyard, Atubeileh, Azhyd, BC Myles, BD2412, Ben-Zin, Benchamas, BerneyBoy, Blainster, Blokenearexeter, Bmackcw, Britzingen, Brockert, Brya, CNWG, CTZMSC3, Cmapm, Connormah, Curtis Clark, DanielCD, Dankarl, Dendodge, Dinoguy2, Dj Capricorn, DocWatson42, Donarreiskoffer, DrMicro, Dustball, Dysmorodrepanis, Easchiff, El C, Eleassar, Eric Kvaalen, Euchiasmus, Eug, Ewen, Famartin, Fir0002, Folypeelarks, Friendlyliz, FrummerThanThou, Gene Nygaard, Geniac, GerardM, Gigemag76, Glenn, Gob Lofa, GorillazFanAdam, Graminophile, Ground Zero, Hardyplants, Hesperian, Imc, J.delanoy, Jaknouse, Jason M, Jauhienij, Jhbdel, Jmgarg1, JoJan, Joel7687, Julienvr, Jóna Þórunn, Kahkonen, Karl-Henner, Kate, Kingdon, Knutux, LOL, Lady Tenar, Lamlam, Lenticel, LilHelpa, Logan, Look2See1, MPF, Marknesbitt, Marshman, Masterthomas, Mgiganteus1, Miaow Miaow, Michaplot, Mike Dallwitz, Miss Madeline, Morwen, MrDarwin, Munita Prasad, NTox, Ngstanton, Nipisiquit, Nono64, NotTires, Nurg, OlEnglish, Paalexan, Peanutcactus, Pedro Onativia, Pekinensis, Phatom87, PierreAbbat, Piolinfax, Pion, Quebec99, Qwertzy2, RandomP, Redvers, Renato Caniatti, RepublicanJacobite, Richard New Forest, RichardMills65, Rob Hooft, Robin S, Rosser1954, RoyBoy, Shattered Gnome, Sin-man, Skizzik, Sluzzelin, Smartse, Smith609, SoilMan2007, Sonelle, Stan Shebs, Stevertigo, Storkk, Tau'olunga, Template namespace initialisation script, Think outside the box, Tim1357, Tjunier, Toby Bartels, Tony Sidaway, UtherSRG, Vina, Viscious81, Vuong Ngan Ha, Watcharakorn, Weedgarden, Wetman, Weyes, Wysprgr2005, Youssefsan, Zeamays, 124 anonymous edits

Poales *Source*: http://en.wikipedia.org/w/index.php?title=Poales *Contributors*: Alan Liefting, Altenmann, Arch dude, Azhyd, Bjankuloski06en, Brya, Calliopejen1, CanisRufus, Chochopk, Conversion script, DanielCD, Dinoguy2, Dj Capricorn, Dysmorodrepanis, EncycloPetey, Glenn, Hesperian, Imc, In actu, Iorsh, Jaknouse, Jmgarg1, JoJan, Josh Grosse, Kazubon, LiDaobing, LilHelpa, Lockesdonkey, Look2See1, MPF, Marshman, MrDarwin, Naddy, Nk, NoahElhardt, Notafish, PDH, Psychless, Rolf Schmidt, Stan Shebs, Stemonitis, TeunSpaans, Toby Bartels, Tom Radulovich, Vuong Ngan Ha, Wlodzimierz, Youssefsan, نیپ لاه, 18 anonymous edits

Commelinids *Source*: http://en.wikipedia.org/w/index.php?title=Commelinids *Contributors*: 1978, B1iw, Betacommand, Bjankuloski06en, Borgx, Brya, Calimo, Dtobias, Dysmorodrepanis, Eichhornia, EncycloPetey, Fanghong, Hesperian, Hqb, Jmgarg1, Kingdon, Look2See1, MPF, Plantsurfer, Rkitko, Stemonitis, Syp, Vuong Ngan Ha, 7 anonymous edits

Monocotyledon *Source*: http://en.wikipedia.org/w/index.php?title=Monocotyledon *Contributors*: 1978, Ahoerstemeier, Alansohn, Archenzo, Atif.t2, Azhyd, Banana04131, Bjr35, Bluelion, Bobo192, Bomac, Brighterorange, Bruce1ee, Brya, CalicoCatLover, CambridgeBayWeather, Candorwien, Carmichael, Ceyockey, Christian75, Circeus, Cmccomb, Comppsi, CopperKettle, Crazymonkey1123, Cropscience2, CyberSyn, DA3N, Daderot, Dany1217, DocWatson42, Dragoon0515, Dungodung, Dysmorodrepanis, E rulez, Elipongo, EncycloPetey, Excirial, Fauzan Zaid, Figma, Flakinho, Folypeelarks, France3470, GerardM, Hajatvrc, Hans Dunkelberg, Hardyplants, HarryH2O, Hbaryan, Henning Makholm, Hesperian, Hippophaë, Hon371, Hult041956, Hypehuman, Imc, Intersofia, Iorsh, Iph, Iquseruniv, Izodman2012, JForget, Jaknouse, JamesLee, Jauhienij, Jefffire, Jewk, Jitse Niesen, Jmgarg1, JoJan, Josh Grosse, Julia W, Jurema Oliveira, Just Another Dan, Jynus, Karl-Henner, Kazrak, Kingdon, Kleopatra, Koven.rm, Lavateraguy, Leflyman, Lightdarkness, Looxix, Lumos3, MPF, Madhero88, Mani1, Marshman, Matt Deres, Mav, Medeis, Meggar, Mgiganteus1, Million Moments, Morfy, Moshe Constantine Hassan Al-Silverburg, MrDarcy, MrDarwin, Munita Prasad, NJPharris, Namar170, NellieBly, Nk, OlEnglish, Ongjyhseng, Oreo Priest, PDH, Pengo, Peter coxhead, Pgan002, Pharaoh of the Wizards, Pinethicket, PuzzletChung, Qxz, Qzd800, Railer 896, ResearchRave, Rjwilmsi, RoRo, Robert Foley, S3000, Salvio giuliano, Semolo75, Seren-dipper, Skwerl masta, Sortior, Starx, Stemonitis, Stepnwolf, Template namespace initialisation script, The Cunctator, The Thing That Should Not Be, TheAlphaWolf, Tigershrike, Tom Radulovich, Tom.russell25, Tomer T, Twinsday, Tycho, Uncle Dick, Unyoyega, Vanished user 03, Vina, Viswaprabha, Vuong Ngan Ha, Wknight94, WriterHound, Youssefsan, Zephalis, Zigger, 218 anonymous edits

Flowering plant *Source*: http://en.wikipedia.org/w/index.php?title=Flowering_plant *Contributors*: (, 2004-12-29T22:45Z, 4twenty42o, 78.26, AUG, Abdul raja, Acdx, Adashiel, Addshore, Addthesmell, Agong1, Alai, Alan Liefting, Alansohn, Alexandra lb, Alexei Kouprianov, Alfirin, Allstarecho, Alnokta, Andreworkney, Andycjp, Anomalocaris, Antandrus, AnthonyBryars, AquamarineOnion, Aranea Mortem, Arcadian, Arch dude, Arkuat, Arthena, AtticusX, Avicennasis, Avono, AxelBoldt, Azcolvin429, Azhyd, Backslash Forwardslash, Bakudai, Baralum, Bballmaster024, Bbplayer4, Bdiscoe, Before My Ken, Ben Hanke, Bender235, Berton, Betterusername, Blockwizzard, Bob the Wikipedian, Bobo192, Brian Crawford, Brianga, Brooke87, BrownHairedGirl, Bruce Marlin, BrunoRreategui, Brya, Bryan Derksen, Bunchofgrapes, Buthwiki, CSWarren, Caltas, Calvin 1998, CambridgeBayWeather, Camw, Can't sleep, clown will eat me, Capricorn42, CarolSpears, Cashew2, Caster23, Cayman20, Cbrodersen, Cflm001, Cheesetuna, Chkenboy, Chrishy, Chrisjj, Chun-hian, Closedmouth, Cnilep, Codiferous, Colby Farrington, Colchicum, Columba livia, Conversion script, Copsandrobbers, Courcelles, Cripdyke, Cureden, Curtis Clark, Cxz111, Cyrius, DARTH SIDIOUS 2, Dan Gluck, Dan Koehl, Darth Panda, Dcoetzee, DeadEyeArrow, Deagle AP, Denisarona, Der Golem, DerHexer, Determinate, Dewert, Diannaa, Dinoguy2, Discospinster, Dj Capricorn, Dmacauley, DomenicDenicola, Don't do this HardyPlants, DrFO.Jr.Tn, DrKiernan, DrMicro, Dragonaxl, Drphilharmonic, Drt1245, Dysmorodrepanis, E smith2000, E2m, EDUCA33E, ELefty, EamonnPKeane, Earthdirt, Eassin, Eclecticology, EconoPhysicist, Eichhornia, ElKevbo, Eleassar, EncycloPetey, Epbr123, Eric Forste, Eric Kvaalen, Etxrge, Euryalus, FCSundae, Fabio99999, FetchcommsAWB, Fillard, Fingerz, Fredgoat, Frogger19961, Furrfu, Gabbe, Gail, Gaius Cornelius, Galoubet, Garcsera82, GaryColemanFan, Geezerbill, Geonarva, Giftlite, Ginger1112, Gioto, Glacialfox, Glenn, Globalphilosophy, Gogo Dodo, GolbatTheSexy, GoneAwayNowAndRetired, Graminophile, Grammarmonger, Gscshoyru, Gwandoya, HJ Mitchell, Haabet, Hadrianheugh, Halcon3324, Hardyplants, Haza-w, Hdt83, Hectorthebat, Heegoop, Helikophis, Heman, Hemmer, Henrikhenrik, Hesperian, HighKing, Hinakana, Hive001, Hopefulromntic, Hulu123456789, Hydrogen Iodide, HyperNovaVII, I hate skool-- --, II MusLiM HyBRiD II, Immunize, Iorsh, Iridescent, Island Monkey, Izybella, J. Spencer, J.H.McDonnell, J.delanoy, JNW, Jackhynes, Jager123, Jaknouse, Jeepday, JerrySteal, Jhbdel, Jim1138, Jmgarg1, Jneves121, JoJan, Joe Jarvis, John254, Jonathan Hall, Jonhen8, Jonik, Joseph Solis in Australia, Josh Grosse, Jrtayloriv, Jujutacular, Jusdafax, Just Another Dan, Karl-Henner, Kazvorpal, Killdevil, Killerrabbitsontheloose, KimberleyReid, Kingdon, Kingpin13, Kingturtle, Kitachi, Kitkatcrazy, Kleopatra, KokomoNYC, Komahadara666, Kouka, Krl, Kuru, KyraVixen, L Kensington, Lavateraguy, Leana222222222, LeaveSleaves, LedgendGamer, Lenticel, Li4kata, Liamdaly620, Liene, Lights, Lockesdonkey, Logan, MPF, Macedoniarulez, Magiccast, Mani1, Manop, Marshman, Matthias M., Mav, McSly, Menchi, Mendaliv, Mephistophelian, Mercury, Mgiganteus1, Michaplot, Mike Dallwitz, Mike Rosoft, Mike and taye, MikeVitale, Mikems, Millahnna, Mitsukai, Mivf, Mmaaddiiee, Mmm, Mollymolly9, Moon&Nature, Morgan Wick, Mowgli, Mozzyepic24, MrDarwin, Mrgreen159, Munita Prasad, NJPharris, NTox, Netoholic, Nickj, Nickname97, Nihil novi, NoahElhardt, Nolaman712, Northamerica1000, Notheruser, Nsaa, NuclearWarfare, Orphan Wiki, PDH, PGWG, Pagingmrherman, PaleCloudedWhite, Pcbene, Pekinensis, Pengo, Penusofnazarethpart2, Peter coxhead, Pethan, Phantomsteve, Pharaoh of the Wizards, PhilKnight, Philip Trueman, Philopp, PierreAbbat, Pietdesomere, Pikachuwashere, Planet plant, PlatypeanArchcow, Polinizador, Pollinator, Polymerbringer, Poor Yorick, Pseudofusulina, Punarbhava, Purplellama31, PuzzletChung, Quandaryus, Qwyrxian, RAW65, Reedy, Renata, RexNL, Richard Keatinge, Richard001, Richardcavell, Rickproser, Ritiksac, Rjwilmsi, Robert Foley, Romanski, Romeslayer, RoyBoy, Rrburke, Rror, Rsrikanth05, RxS, Sasata, SchfiftyThree, Schwanzlutscher Ficksau Dauergeil, Scilit, Sciurinæ, Seaphoto, Secretlondon, Seglea, Sfdan, Shalom Yechiel, Shannon1 (usurped3), Sky Attacker, Slon02, Smith609, Snigbrook, SoCalSuperEagle, Sortior, SpeedyGonsales, Srushe, Stanhopea, Stemonitis, Sterio, Steven Zhang, Suffusion of Yellow, Sugarcoated817, Superm401, Svetovid, TUF-KAT, TaraBartolec, Tau'olunga, Tcncv, Template namespace initialisation script, The High Fin Sperm Whale, The Thing That Should Not Be, TheodorCH, Thepcnerd, Tide rolls, Tim Ross, Tinymouth, Tom Radulovich, Tom harrison, TomPhil, Tomfy, Tomi, Toroca, Total Tom, Tpbradbury, Transope, Traveler100, Triona, Trusilver, Ufwuct, Uncle Dick, User A1, UtherSRG, Vald, Vanderesch, Vanished user, VashiDonsk, Veddharta, Versageek, Versus22, Vetoeyou, Vlmastra, Voxii, Vrenator, Vsmith, Vuong Ngan Ha, Vzb83, Waggers, Wikimachine, Willking1979, Wimt, Wingchi, Wknight94, Wlodzimierz, Woohookitty, Wotflowers, Writtenonsand, Wtmitchell, X!, XJamRastafire, Xndr, Xskipx7777, YebisYa, Yosri, Yoyospaghettio, Zanimum, Zeamays, Zeman, Zidonuke, Ziegelangerer, , 749 anonymous edits

Image Sources, Licenses and Contributors

Image:NSRW Auxanometer.png *Source*: http://en.wikipedia.org/w/index.php?title=File:NSRW_Auxanometer.png *License*: unknown *Contributors*: The New Student's Reference Work

File:Asteracea poster 3.jpg *Source*: http://en.wikipedia.org/w/index.php?title=File:Asteracea_poster_3.jpg *License*: unknown *Contributors*: User:Alvesgaspar, User:Tony Wills

Image:Stem-histology-cross-section-tag.svg *Source*: http://en.wikipedia.org/w/index.php?title=File:Stem-histology-cross-section-tag.svg *License*: unknown *Contributors*: User:Emmanuel.boutet

Image:spikes.jpg *Source*: http://en.wikipedia.org/w/index.php?title=File:Spikes.jpg *License*: unknown *Contributors*: Hardyplants

File:Angiosperm life cycle diagram.svg *Source*: http://en.wikipedia.org/w/index.php?title=File:Angiosperm_life_cycle_diagram.svg *License*: unknown *Contributors*: user:LadyofHats

Image:Aesculus hippocastanum fruit.jpg *Source*: http://en.wikipedia.org/w/index.php?title=File:Aesculus_hippocastanum_fruit.jpg *License*: unknown *Contributors*: Solipsist

Printed by Books on Demand GmbH, Norderstedt / Germany